Jay V. Huner, PhD
Editor

Freshwater Crayfish Aquaculture in North America, Europe, and Australia: Families Astacidae, Cambaridae, and Parastacidae

Pre-publication
REVIEWS,
COMMENTARIES,
EVALUATIONS . . .

" **T**he book *Freshwater Crayfish Aquaculture in North America, Europe, and Australia* brings together some of the top crayfish scientists from around the world. The individual chapters are well organized and one chapter flows to the next. THIS BOOK PRESENTS THE BEST AND LATEST OVERVIEW OF CRAY-FISH CULTURE ON A WORLD BASIS, AND IT WILL ADD IMMEA-SURABLY TO OUR KNOWLEDGE BASE."

James W. Avault, Jr., PhD
School of Forestry, Wildlife, & Fisheries, Louisiana State University

More pre-publication
REVIEWS, COMMENTARIES, EVALUATIONS . . .

" **I** found *Freshwater Crayfish Aquaculture in North America, Europe, and Australia* to be an excellent compilation of information regarding crayfish. This book covers much more than crayfish culture. IT IS AN UP-TO-DATE, COMPREHENSIVE TREATISE OF CRAYFISH USEFUL TO ANYONE INTERESTED IN CRAYFISH RESEARCH, BIOLOGY, CULTURE, HARVEST, MARKETING, AND PROCESSING. I appreciated the inclusion of more basic crayfish biology and behavior information which is important to fully understand crayfish and their aquaculture potential.

As an aquaculture extension educator, I frequently get requests for information from people interested in all aspects of crayfish culture. This book will be very helpful in responding to those requests.

I wish I would have had a resource like this available to me when I first became involved with crayfish. If you are interested in culturing, harvesting, or marketing crayfish, there is no sense reinventing the wheel or wasting time and money on techniques that have not proven successful. This comprehensive discourse on crayfish culture and related topics, from three continents, describes the state of the art and points to areas that need further study. The book is also valuable to researchers for its extensive review of the literature. I HIGHLY RECOMMEND THIS BOOK."

Jeffrey L. Gunderson, MS, BS
Assistant Specialist–Fisheries,
Sea Grant Extension Program
and the Natural Resources
Research Institute,
University of Minnesota

" ***F****reshwater Crayfish Aquaculture in North America, Europe, and Australia* will be a welcome addition to the library of anyone interested in aquaculture. IT WILL BE A MUST FOR THOSE INTERESTED IN CRAYFISH BIOLOGY, CULTURE OR PROCESSING AND MARKETING. The book is divided into three sections, each on a different region of the world and each written by recognized scientists in the region.

The book is well-written, provides a good review of the current literature with references for further reading, and has a great deal of previously unpublished information offered from the experiences of the various contributors."

David B. Rouse, PhD
Associate Professor,
Department of Fisheries,
Auburn University

Food Products Press
An Imprint of The Haworth Press, Inc.

Freshwater Crayfish Aquaculture in North America, Europe, and Australia
Families Astacidae, Cambaridae, and Parastacidae

Freshwater Crayfish Aquaculture in North America, Europe, and Australia
Families Astacidae, Cambaridae, and Parastacidae

Jay V. Huner, PhD
Editor

Food Products Press
An Imprint of The Haworth Press, Inc.
New York • London • Norwood (Australia)

Published by

Food Products Press, an imprint of The Haworth Press, Inc., 10 Alice Street, Binghamton, NY
13904-1580

Freshwater crayfish aquaculture in North America, Europe and Australia : families Astacidae,
Cambaridae, and Parastacidae / Jay V. Huner, editor.
     p.   cm.
     Includes bibliographical references and index.
     ISBN 1-56022-039-2 (acid free paper).
     1. Crayfish culture. 2. Astacidae–Breeding. 3. Cambaridae–Breeding. 4. Parastacidae–
Breeding. I. Huner, Jay V.
SH380.9.F74   1993
639'541–dc20   92–21938
                                                 CIP

CONTENTS

ABOUT THE EDITOR

Jay Huner, PhD, is Director of the Crawfish Research Center and Adjunct Professor of Aquaculture at the College of Applied Life Sciences at the University of Southwestern Louisiana in Lafayette. A Certified Fisheries Scientist, he has extensive teaching, research, and editorial experience in the field. Dr. Huner is the author of over one hundred technical and semi-technical fisheries publications and has co-authored or co-edited several aquaculture-related books. He is General Manager for the International Association of Astacology (IAA), a member of the American Fisheries Society, the World Aquaculture Society, and the National Shellfisheries Association, and is active in many other professional organizations. Dr. Huner is editorially involved with the IAA Newsletter, *Farm Pond Harvest Magazine, Small Farm Today Magazine*, and the Louisiana Soft-Shell Crawfish Association Newsletter.

List of Figures

North American Section

European Section

Australian Section

List of Tables

Introduction

This text has been written to summarize the present state of knowledge about freshwater crayfish aquaculture.[1] Crayfishes represent the dominant, keystone benthic invertebrates in many temperate aquatic ecosystems in North America, South America, Europe, Eastern Asia, Australia, and New Guinea. Several parastacid species are truly tropical animals with ranges including northern Australia, New Guinea, and Indonesia. Crayfishes are, however, largely absent from the tropics except for one cambarid species, *Procambarus clarkii*. This species, commonly called the red swamp crayfish, has been introduced from the southern USA into Central and South America, Africa, the Middle East, western Europe, southern Asia, and Hawaii.

Crayfishes are exploited through natural fisheries wherever they are found but important commercial, recreational, and/or subsistence fisheries are present only in the USA, Australia, New Guinea, Kenya, Spain, Scandinavia, Turkey, and mainland China. Significant commercial aquaculture is limited largely to the southern USA but considerable interest in their culture, as evidenced by numerous, small scale aquaculture endeavors, is seen elsewhere in the USA, Europe, and Australia.

Total annual commercial harvest of freshwater crayfishes exceeds 110,000 metric tonnes with 55% coming from the USA, primarily in the southern state of Louisiana, 36% from the People's Republic of China, 8% from Europe, and less than 2% from Australia. The cosmopolitan red swamp crayfish accounts for 70-80% of this total. Potential for increases in production of this species through aquaculture and natural fisheries is great.

---

1. The word crayfish is used in favor of the word crawfish because it is a more cosmopolitan term. It includes those freshwater crustaceans belonging to the families Astacidae, Cambaridae, and Parastacidae.

This text does not seek to provide an in-depth treatment of cray-fish biology, physiology, taxonomy, and zoogeography except as it pertains to crayfish culture. Numerous texts referenced in this book may be consulted for that information. Rather, it is our intention to provide the reader with one source of information about freshwater crayfish aquaculture. This, we feel, will provide the reader with methods employed around the world. These differ widely according to the species concerned and varying technical problems encountered around the globe.

We believe that the reader should be able to synthesize various methods to develop workable strategies for cultivating different species in specific environments. For example, hatchery technology has been important in Australia and Europe but has yet to be applied to any degree in North America. As a result, a review of North American crayfish culture would, necessarily, have little to say about the subject. However, the huge volume of crayfish harvested in North America has led to the development of a major crayfish processing industry there, principally in the state of Louisiana. As a result, a review of North American crayfish culture will include a detailed discussion of this subject that would not be found in reviews of Australian and European crayfish culture. Furthermore, the significance of the crayfish fungus, *Aphanomyces astaci*, which is the organism responsible for the disease that has decimated native astacid crayfishes in Europe for over a century, would not be obvious in a review of North American crayfish cultural methods. This is because North American species are highly resistant to the malady. In addition, Australian species are believed to be highly susceptible to it. It has not yet been reported in any parastacid populations in their native ranges.

Fewer than a dozen crayfish species are cultivated and fewer than two dozen species sustain important commercial fisheries. A number of the 300 plus cambarid species do, however, have aquaculture potential as do several of the 90-100 parastacid species.

We must emphasize that we do not advocate the introductions of any crayfish species outside of its native range. Decisions to make such introductions should be made only by representative governmental authorities in consultation with scientists, including crayfish specialists, or astacologists.[2] It must be assumed by anyone contemplating the aquaculture of a species outside of its native range that successful culture will invariably lead to a successful introduction whether or not the project is a commercial success.

We hope that this text will prove valuable and worthwhile to the reader. We welcome comments, negative and positive, so that we can improve future editions.

Organizer of this book is Jay Huner. Authors include, in alphabetical order: Hans Ackefors (Sweden), Jay Huner (USA), Ossi Lindqvist (Finland), B. J. Mills (Australia), Michael Moody (USA), Noel Morrissy (Australia), and Ronald Thune (USA).

The authors wish to thank all individuals who assisted in the production of this manuscript. Special thanks are extended to Ms. Janice Meaux and Ms. Julie Mills for their secretarial assistance in producing the final manuscript.

---

2. The term astacology was coined in 1972 when the International Association of Astacology was founded in Hinterthal, Austria. It is taken from the scientific name for the first crayfish *Astacus astacus* described by the Swedish taxonomist Carolus Linnaeus, Astacidae. The permanent Secretariat for the International Association of Astacology is located in Lafayette, Louisiana (USA). The address is: P.O. Box 44650, University of Southwestern Louisiana, Lafayette, Louisiana 70504-4650, USA.

Cultivation of Freshwater Crayfishes in North America

Jay V. Huner, Section I

Crawfish Research Center
University of Southwestern Louisiana
Lafayette, Louisiana, USA

Michael Moody, Section II

Louisiana Cooperative Extension Service
Louisiana State University
Baton Rouge, Louisiana, USA

Ronald Thune, Section III

School of Veterinary Medicine
Louisiana State University
Baton Rouge, Louisiana, USA

Section I:
Freshwater Crayfish Culture

INTRODUCTION

North America is second to no other land mass in its freshwater crayfish fauna. It includes two families, Astacidae and Cambaridae, 12 genera, *Pacifastacus* (Astacidae) and *Barbicambarus*, *Bouchardina*, *Cambarus*, *Cambarellus*, *Distocambarus*, *Fallicambarus*, *Faxonella*, *Hobbseus*, *Orconectes*, *Procambarus*, and *Troglocambarus* (Cambaridae), and well over 300 species (Hobbs 1988). The bulk of

5

the species belong to the family Cambaridae and are naturally distributed east of the great western mountains that divide the continent. However, one *Orconectes* species has naturally crossed that barrier and two other aggressive, large species, *Orconectes virilis* and *Procambarus clarkii*, have been successfully introduced (Bouchard 1978; Hobbs et al. 1989). The genus *Pacifastacus* is the only representative of the family Astacidae in North America. Native to the west coast from British Columbia well into California, at least one representative of the genus, *P. leniusculus*, is a large, aggressive species that has been widely introduced within the region but has yet to be moved eastward within the continent.

The commercially important crayfish species in North America include, in order of significance, the red swamp crayfish, *Procambarus clarkii*, the white river crayfish complex, *Procambarus* spp.,[1] the signal crayfish, *Pacifastacus leniusculus*, the rusty crayfish, *Orconectes rusticus*, the papershell crayfish, *Orconectes immunis*, and the fantail, or northern crayfish, *Orconectes virilis* (Huner 1989c). Other species are harvested when they are locally abundant and some invade fish culture systems. In addition, there are so many other species that more species are sure to be eventually involved in some form of exploitation, either through harvesting of wild populations and/or culture. Most likely candidates will be tertiary burrowers of the genus *Procambarus*, subgenera (*Ortmanicus*) and (*Scapulicambarus*), and larger *Orconectes* spp.

The most important single crayfish species is *P. clarkii* which accounts for 70-80% of all crayfish harvested in North America, natural fisheries and culture combined (Huner 1989c). It is cultivated along with white river crayfish throughout the southern and southeastern USA. The state of Louisiana dominates the industry

---

1. The white river crayfish complex was once identified as one cosmopolitan species, *Procambarus acutus acutus* in the eastern half of the USA. However, Hobbs and Hobbs (1990) have described a new species, *Procambarus zonangulus*, from the western and central rim of the northern Gulf of Mexico. This was previously considered to be *P. a. acutus* (H. H. Hobbs, Jr., 1990, Smithsonian Institution, Washington, DC, USA, personal communication). *Procambarus a. acutus* occurs to the east of *P. zonangulus* to and along the east coast of the U.S.A. Furthermore, an unidentified species, previously identified as *P. a. acutus* occurs throughout the central USA from central Louisiana into Wisconsin. Note: At one time *P. a. acutus* was called *Procambarus blandingi*.

with 50,000 ha of earthen ponds (Sandifer 1988). Information about production in the USA is presented in Table 1. The other important cultured species is *O. immunis*. No survey data are available about culture of this species but pond area devoted to its culture probably approaches 500 ha in states bordering the Great Lakes between Canada and the USA. Reference to its intentional culture for fish bait dates from the early 1940s (Tack 1941). This species was largely considered to be too small for food markets but closer examination has demonstrated that, under the proper conditions, it can be cultivated to much larger sizes (McCartney and Garrett 1989).

The North American crayfishes range in size from the tiny dwarf species of the genus *Cambarellus*, which rarely exceed 1 g to the large members of the genera *Cambarus*, *Orconectes*, *Pacifastacus*, and *Procambarus* which can, under suitable conditions, achieve sizes of 50-80 g. Growth is rapid compared to that observed in European species because climate is much milder. Maturation and reproduction can be realized in one year throughout the continent although this often requires two years at higher latitudes (Payne 1978; Momot 1984).

North American crayfishes have long been used as food, fish bait, specimen for specific research and teaching, pets, and objects of social significance (Huner and Barr 1991). They were first exploited by the continent's original aboriginal inhabitants and later by the multitude of ethnic groups who came later. The first records of crayfish culture date from the late 1700s when Louisiana plantation owners sometimes raised them in small ponds for the family's kitchen (Comeaux 1975). While larger species are preferred for food, even the dwarf crayfishes were and still are collected for food in Mexico. These are eaten whole as a part of taco fillings.

Intentional commercial crayfish culture seems to date from the 1930s and 1940s with references to their culture for fish bait and food appearing in the literature then (Langlois 1935; Viosca 1937; Tack 1941). Crayfish culture for fish bait developed in the Great Lakes states largely as an adjunct to regular finfish cultural endeavors (Huner 1978a). Crayfish culture for food developed principally in the state of Louisiana as an adjunct to the rice industry there (Viosca 1966). However, no determined effort has been made by the Louisiana industry to pursue fish bait markets (Huner 1988b).

Table 1. Crayfish Production (Tons) in the U.S.A. (1990).

Region	Wild	Aquaculture	Species
Southern	5,000–25,000	60,000 (65,000 ha)	*Procambarus clarkii* *P. zonangulus** *P. a. acutus** *Procambarus* sp.*
Northcentral	100	150 (500 ha?)	*P. clarkii* *Procambarus* sp.* *Orconectes immunis* *O. rusticus* *O. virilis* *Orconectes* spp.
Northeastern	?	25	*O. immunis* *Orconectes* spp.
Western	300-500	none	*Pacifastacus leniusculus* *P. clarkii*

* These three species were previously considered to be one species *Procambarus acutus acutus.*

Both forms of crayfish culture share common origins in that crayfish invaded existing fish culture ponds and rice fields. While it is feasible to cultivate crayfish from egg to egg within closed systems, no one has yet been able to make such an endeavor financially successful (Huner 1990d). Soft-shell crayfish are produced profitably in tanks and pools but only by harvesting the animals from earthen ponds where they have been grown in extensive culture to a suitable size before transfer to tanks and pools.

TAXONOMIC CONSIDERATIONS

Taxonomy of the cambarid crayfishes is based on the terminal shape of the male sperm transfer organs, the gonopodia, and the female sperm receptacle, or annulus ventralis (Hobbs 1972, 1981). The gonopodia are the first pair of abdominal appendages which extend anteriorally between the walking legs. The annulus ventralis is located in the sternal area between the bases of the walking legs. These two sets of sexual structures develop very distinctive morphologies when a crayfish matures. A species-specific groove in the annulus ventralis leads to an internal storage area for sperm. Taxonomic keys for describing species have been developed only for male cambarid crayfishes. The gonopodia of male astacid crayfishes do not develop similar terminal morphologies. The female astacid crayfishes do not have seminal receptacles. Rather, spermatophores are attached to the sternal area during the mating process. The reader is referred to various taxonomic treatises by Dr. Horton H. Hobbs, Jr. for more information about identification of North American crayfishes (Hobbs 1972, 1974, 1981, 1989).

BASIC LIFE CYCLE CONSIDERATIONS

North American crayfishes are, as stated earlier, cultivated by establishing sustaining populations in earthen ponds. These ponds are invariably dried intentionally for crayfish management purposes or for fish husbandry purposes with the crayfishes surviving in burrows. Therefore, it is important to understand basic life cycles. (This discussion will be limited to cambarid crayfishes. The life cycles of astacid crayfishes are covered in the European section of this book.)

One North American astacid species, *P. leniusculus,* has been widely distributed throughout western Europe and young are cultivated for establishing sustaining populations in semi-controlled pond units which are not drained (Westman et al. 1990). Astacid crayfishes cannot persist in burrows in dry ponds because they are not physiologically able to live in low oxygen environments.

Female cambarid crayfishes retreat to burrows or secluded areas to lay and incubate eggs (Payne 1978; Huner and Barr 1991) (Figure 1). Eggs exit from openings of the oviducts located at the bases of the third pair of walking legs counting the chelipeds, cross the annulus ventralis to the folded abdomen where they will become attached to the swimmerets, or pleopods (Andrews 1906). Sperm is released from the annulus ventralis and fertilization is external.

Cement, or glair glands located on the underside of the abdomen and tail fan release a mucus-like substance called glair in which the eggs are fertilized and become attached to setal hairs on the swimmerets. The outer wall of the egg provides the attachment membranes, not the glair material. Prior to laying eggs, the glair glands enlarge and extensive areas of white "patches" are apparent on the ventral area of the abdomen, especially on the tail fan.

The incubation period depends on temperature and may be as short as 2 1/2-3 weeks at 22-24°C or as long as 6 months when eggs are laid in the autumn at northern latitudes. Suko (1954, 1956) provides information on egg development of *P. clarkii* at different temperatures, (see Table 2). While this species is normally considered to be a southern, warmwater species, it has adapted well to

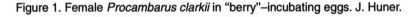

Figure 1. Female *Procambarus clarkii* in "berry"–incubating eggs. J. Huner.

more northerly locations in upper Ohio (Huner 1984) and Japan (Suko 1958).

Once the young crayfish have hatched, they cling to the female's swimmerets and undergo two molts, the duration of time between these molts depends on temperature (Andrews 1907). They are unable to survive away from the female until they have completed their second molt. At that time they assume an adult-like appearance and begin to feed, no longer subsisting on stored egg yolk. Young will remain with the female for several weeks after the second molt in confined quarters, being attracted by a brood pheromone (Little 1975, 1976). Once the female leaves the brooding burrow young crayfish become dislodged and are unable to return to her.

The young crayfish undergo approximately 11 molts altogether before they become sexually mature (Suko 1953; Black 1966). Size at maturity depends upon genetic and environmental constraints (Huner and Romaire 1979; Momot 1984; Lutz and Wolters 1986). The importance of density in determining growth potential, and dominance of mature males in determining survival cannot be overemphasized. Density is clearly far more important than genetics in determining size at maturity.

Table 2. Development of Fertilized *Procambarus clarkii* Eggs as Influenced by Temperature, after Suko (1956).

Period of Development (days)	Mean Temperature (°C)	Degree Days
17	22.8	388
23	21.8	501
29	20.0	580
36	17.9	644
46	15.0	690
59	11.8	696
130	10.0	1300
155	8.9	1381
162	9.0	1428

Mature male and female cambarid crayfishes are described as being in the Form I condition (Hobbs 1972) (Figure 2). They will not molt again before the reproductive cycle is completed. Mature males develop very distinctive secondary sexual characteristics including very inflated/enlarged chelae (claws), cornified gonopodia, and hooks at the bases of the third and/or fourth pairs of walking legs. The chelae of mature females enlarge somewhat and the annulus ventralis cornifies.

Mature cambarid crayfishes mate freely and the female may initiate the mating act (Figure 3). Ameyaw-Akumfi (1981) has described the mating behavior of *P. clarkii*. Males without chelae can mate but cannot successfully compete with males having intact chelae.

Figure 2. *Procambarus clarkii,* left to right, mature, form I male, mature "form I" female, and juvenile male. Note the relative proportions of the chelae with the form I male having the largest chelae. J. Huner.

Once the reproductive cycle has been completed, surviving crayfish feed vigorously and molt into a quasi-juvenile growth stage called the Form II condition (Figure 2). They will subsequently molt, normally back to the sexually active Form I condition. However, they may, in some cases, undergo two or more molts before reverting to the Form I condition (Kossakowski 1966; Price and Payne 1979). Hobbs (1981) describes a method to identify Form II male crayfish by inspection of sutures on limbs. Cambarid crayfishes mature over such a wide size range that size is not a satisfactory way to identify Form II crayfish (Figure 4).

Maturation can take place in about 3 months under ideal environmental conditions. In practice, maturation generally takes 5-6 months at southern latitudes and 10-18 months at more northern latitudes (Payne 1978). Some species like *P. clarkii* and the dwarf crayfish, *Cambarellus shufeldtii*, mature, mate, and reproduce year

Figure 3. *Procambarus clarkii* mating with male on the right. J. Huner.

round in warmer climates although the majority seem to produce young in the late summer to mid-autumn period (Penn 1943; Black 1966). Most species are annual spawners including the commercially important species besides *P. clarkii* (Payne 1978).

Suko (1953, 1954, 1955, 1956, 1958, 1961) conducted intensive studies of the *P. clarkii* reproductive cycle, development, and growth. These papers should be consulted by any serious cambarid crayfish culturist.

BURROWS

An attractive feature of crayfish aquaculture is the fact that populations are self-perpetuating so that hatchery systems are not re-

Figure 4. The size variation in mature, non-growing form I male *Procambarus clarkii*. The range is about 7-12 cm total length. J. Huner.

quired. The important cultivated species including *P. clarkii,* the white river crayfish complex of species, and *O. immunis* are burrowing species. All thrive in unstable aquatic habitats that are subject to annual or more frequent dewatering and the various *Procambarus* spp. are cultivated by intentionally dewatering ponds for several months during the late spring/early autumn period (Huner 1990d). Thus, the importance of burrows to successful crayfish cultivation cannot be overemphasized.

Hobbs (1981) classifies burrowing crayfishes in three broad categories based on burrow complexity and dependence on the burrow for sustaining the species. Cultivated species are classified as tertiary burrowers. These construct the least complex of all burrow types. The burrows rarely descend more than a meter and are relatively simple tubes. The crayfish excavate or retreat to pre-existing burrows when surface waters disappear, and/or when they are engaged in reproductive activities, and/or are seeking cover. Other commercially important species inhabit permanent habitats. They can excavate very simple burrows but are unable to survive long periods if surface waters disappear (Berrill and Chenoweth 1981).

The secondary and primary burrow classifications refer to species that construct very elaborate burrows and rarely appear at the surface. They are morphologically adapted to a more or less permanent fossorial existence (Hobbs 1975, 1981). None appears to have any cultural potential, at least for food or fish bait (Huner 1990d).

The burrow, then, is a key feature in crayfish cultivation but surprisingly little is known about conditions in burrows and the effect of burrow conditions on crayfish survival and reproduction. Descriptions of water quality for *P. clarkii* burrows in southern Louisiana strongly suggest that water in those burrows serves only to maintain 100% humidity in the system (Jaspers and Avault 1969). Dissolved oxygen levels are generally too low to permit normal submerged respiration forcing the crayfish to utilize atmospheric oxygen within the burrow chamber itself. Presumably, carbon dioxide and nitrogen waste products can be released within free water in the burrow. However, if there is no free water in the burrow, the crayfish is unable to eliminate these waste products in their simple forms. They can continue to utilize atmospheric oxygen as long as their gill chambers are wet and soil moisture can maintain 100% humidity within the burrow (Huner and Barr 1991).

McMahon and Stuart (1990; B. R. McMahon, Biology Department, University of Calgary, Calgary, Canada, personal communication) suggests that species such as *P. clarkii* may actually produce a crystalline nitrogen waste product such as uric acid under such conditions. McMahon (1986) discusses physiological adaptations of crayfishes to survive outside of the aquatic environment and has only recently begun to describe the specific mechanisms found in hardy species such as *P. clarkii* (McMahon and Stuart 1990). Huner (1989a) has observed that adults of both *P. clarkii* and *P. zonangulus* can survive 3-4 months in simulated burrow conditions with 100% humidity but no free water. Juveniles rarely survived more than one month under similar conditions. They can and do burrow in suitable substrate (Rogers and Huner 1985).

It remains unclear whether crayfish can lay and incubate eggs when there is no free water in a burrow. Craft (1980; B. Craft, U.S. Soil Conservation Service, Alexandria, Louisiana, USA, personal communication) found healthy *P. clarkii* in humid burrows without freestanding water during September in a pond shortly before it was filled with water. None of the females were carrying eggs or young but the pond subsequently produced an excellent crop in excess of 1000 kg/ha. The burrows were "dry" for 6-8 weeks.

The inability to see and monitor crayfish broodstock in burrows is a critical problem in crayfish culture as it is now practiced in North America. Research in this area is sorely needed. Additional topics of interest include burrowing success and condition of brood crayfish in the burrows. If adequate numbers of healthy brood crayfish do not burrow, survive through a quiescent period, summer in the South and winter in the North, and produce healthy broods of young, pond failure is assured. Transfer of young crayfish from a secondary source is not yet an economically viable alternative.

The current practice of stocking brood crayfish by weight without close attention to burrowing success and survival leaves much to be desired. This procedure makes crop forecasting prior to pond filling impossible. Furthermore, dependence on surviving, unharvested crayfish to perpetuate pond production in future years is certainly a principal factor in explaining widespread variation in production within the same pond and between ponds, even those adjacent to each other. In fact, it is certainly a wonder that such a huge crayfish industry has developed, principally in Louisiana, under this particular management condition.

MOLTING

Crayfishes like all other arthropods molt periodically in order to grow (Aiken and Waddy 1987). The molt cycle is generally divided into five stages, A–soft, B–postmolt, C–intermolt, D–premolt, and E–the molt itself. This series of general stages can be further divided into more precisely divided subunits which include, A_1–A_2, soft; B_1, B_2, C_1, C_2,–C_3, postmolt; C_4, intermolt; D_0, D_1, D_2, D_3,–D_4, premolt; and E_1, E_2, E_3,–E_4, the molt itself. The various substages correlate with changes taking place in the exoskeleton as the new one is cyclically formed and the old one degenerates. These can be followed precisely by microscopic examination of setae (Stevenson 1975; Huner and Avault 1976a; Aiken and Waddy 1987; Huner and Barr 1991) and less precisely by macroscopic examination of changes in relative hardness and color (Stevenson 1975; Culley and Duobinis-Gray 1990; Huner and Barr 1991).

Premolt changes are of interest to crayfish culturists from the standpoint of production of soft-shell crayfish. As the molt approaches, the old exoskeleton demineralizes and becomes more and more brittle. The new, unmineralized exoskeleton forms beneath it and midway through the premolt stage separates from the old one. At that time one can break the old exoskeleton and the new unmineralized one will appear beneath it. Body color becomes progressively darker as the heavily pigmented, new epidermis becomes more visible through the old exoskeleton as it demineralizes and becomes increasingly translucent.

All freshwater crayfishes store some calcium solubilized from the old exoskeleton in paired hemispherical stones located adjacent to the stomach (Huner and Barr 1991). These stomach stones, or gastroliths, are molted with the unmineralized lining of the stomach into the lumen of the stomach where they are dissolved. The calcium is then absorbed into the circulatory system and redeposited into the new exoskeleton. However, calcium stores in the gastroliths as well as the hemolymph and the hepatopancreas are only adequate for the crayfish to mineralize its exoskeleton to about one third the intermolt level (Huner et al. 1978). Remaining calcium must come from food and/or the water.

Stomach stones should be removed from soft-shell crayfish before they are served as food. If they are not removed, the consumer could damage a tooth when eating a soft-shell crayfish (Huner 1988c; Culley and Duobinis-Gray 1990).

Molting is a cyclic phenomenon controlled by hormones from the paired medulla terminalis x-organs located in the eyestalks and the paired y-organs located in the body proper just behind the eyestalks (Aiken and Waddy 1987). The x-organ hormone is a peptide that inhibits molting and is called molt-inhibiting hormone, MIH. The y-organ hormone is a steroid that stimulates molting. It is often called molt-stimulating hormone, MSH, but a better name might be molting hormone, MH, to distinguish it from melanocyte stimulating hormone. It is secreted in an inactive form called alpha ecdysone and is activated into beta ecdysone, or ecdysterone, in the body. MIH and MSH act antagonistically. When there are high levels of MIH, molt is inhibited, and conversely molting is induced when there are high titers of MSH.

Crustacean endocrinologists have long known that bilateral eyestalk ablation (removal) would stimulate molting (Aiken and Waddy 1987). Injection of purified phytoecdysones will also induce molting but the molting is normally incomplete unless the hormone is injected in relatively small doses over time.

No one has yet developed an effective procedure to stimulate molting on a practical basis by injecting into or feeding hormones to crayfishes or by performing bilateral eyestalk ablation. Ablation research is being conducted and shows promise (Huner et al 1990).

A juvenile hormone, JH, has been identified in crayfish (Laufer et al. 1986; Borst and Tsukimura 1990). This compound is called methyl farnesoate. It is synthesized by paired mandibular glands located near the mandibles in the body proper. JH functions in regulating metamorphosis from the juvenile and/or quasi-juvenile Form II condition to the mature, Form I condition. JH also serves as a regulator of ovarian development in females. This hormone has been little studied, having only recently been identified, and its functions in crayfishes are not fully understood.

The JH presumably effects secondary sexual characters in males through the androgenic glands. Taketomi et al. (1990) discuss the androgenic gland in *P. clarkii.*

CULTURE

This section will emphasize *Procambarus clarkii* and *Procambarus zonangulus* because they are the dominant species in the North American crayfish culture industry (Figure 5). It will be based on the Louisiana model but comment will be made, where appropriate, as to how crayfish are cultivated elsewhere on the continent.

Pond Types

Most specialists categorize crayfish ponds as open, wooded/ semi-wooded, and ricefield ponds (de la Bretonne and Romaire 1989b; Huner and Barr 1991). Open ponds are normally constructed on cleared, often marginal agricultural lands and are permanently committed to crayfish culture (Figure 6). They are sometimes called permanent ponds to distinguish them from semi-permanent ricefield ponds. Wooded ponds are typically built in swampy areas by surrounding the areas with levees and controlling water levels. Semi-wooded ponds are simply those that have been cleared to some degree to facilitate growth of forage crops, harvesting and water circulation. Ricefield ponds are ricefields used to cultivate crayfish in rotation with agronomic crops including rice, soybeans, and sorghums. Marshy areas are sometimes leveed for crayfish production and such ponds are called marsh ponds.

Pond sizes vary with wooded ponds being largest, sometimes exceeding 50 ha (de la Bretonne and Romaire 1989b). Open ponds and ricefield ponds are usually 8-16 ha. Information about average pond size by type is presented in Table 3.

Production levels from various pond types in Louisiana are also presented in Table 3. Note that while production can reach 3000 or more kg/ha/yr (Huner and Barr 1991), the state's long-term average

Figure 5. Form I male red swamp crayfish, *Procambarus clarkii* (left), and white river crayfish, *Procambarus zonangulus* (right). Note that the chelae of *P. clarkii* are wider than those of *P. zonangulus.* J. E. Barr.

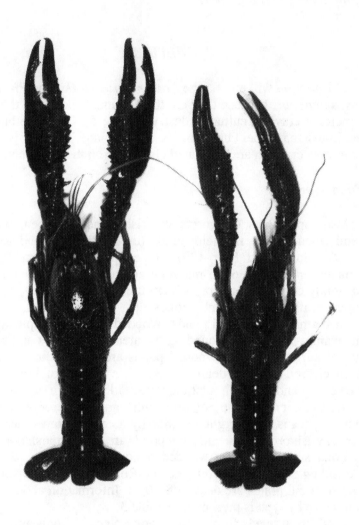

Figure 6. Open crayfish pond planted with rice as forage. J. Huner.

has been 500-600 kg/ha/yr (K. Roberts, 1990, Louisiana Cooperative Extension Service, Louisiana State University, Baton Rouge, Louisiana, USA, personal communication). Similar results are reported from South Carolina (Eversole and Pomeroy 1989) where good production data are maintained and are apparently true for other states where red swamp and white river crayfishes are cultivated.

Site Selection and Construction

Heavy clay and silty clay soils are recommended for crayfish pond construction (Craft 1980; Broussard 1984). Soils that do not retain water must be avoided. At this point, there are no specific guidelines about soil fertility, water table considerations, and so forth. Certainly, sites located where there are pre-existing populations of commercial crayfish species are more likely to be successful than those sites without them (Figure 7). Areas with high levels

Table 3. Louisiana Crayfish Ponds and Production Levels, after de la Bretonne and Romaire (1989b).

Pond Type	Size (ha)	Production* (kg/ha)
Wooded	20-120	560 (90-900)
Semi-Wooded	20-120	670 (225-1,100)
Open		
Permanent	8-40	1,200 (340-3,240)
Ricefield	6-16	1,000 (450-2,800)
Marsh	20-80	450 (200-600)

*Annual production in Louisiana averages 500-600 kg/ha when all pond systems are considered. Many factors, especially economic ones, contribute to this situation.

of residual, long-lived chlorinated hydrocarbon insecticides such as DDT, endrin, aldrin, dieldrin, etc. should be avoided as these substances are highly toxic to crayfishes (Avault and Huner 1985; Huner and Barr 1991).

An important consideration in selecting a pond site is whether or not it is classified as a "wetland." If this is the case, it may be impossible to obtain a permit to construct the pond(s). Failure to consult competent authorities before initiating a construction project can lead to criminal liability! Advisory services such as the U.S. Soil Conservation Service and the local state cooperative extension service should be contacted prior to site selection to ascertain the current status of agricultural construction regulations for the region. Furthermore, both agencies provide free consulting services for siting and constructing ponds.

Crayfish ponds require levees that will hold approximately 0.5 m of water when full with an equivalent height for "free board" to compensate for erosion from wave action (Craft 1980; Broussard 1984; Baker 1987; de la Bretonne and Romaire 1989b; Huner and Barr 1991). Inner and outer levees should be wide enough to permit vehicular traffic and be planted with cover grasses to prevent erosion. All weather access to pump units and boat loading/unloading areas is important. Well-built wide levees help to insure that damage done by burrowing crayfish, mammals, and reptiles can be detected and corrected without levee failure and pond loss. Inner guide, or baffle, levees should be constructed at 50-75 m intervals to insure that water may be thoroughly flushed, leaving no anoxic areas (Figure 8). These levees need only extend above the surface and be wide enough to withstand erosion from wave action during the season. If borrow ditches must be made to provide fill for main levees, they should be located on the outside rather than the inside of levees. Water will follow the path of least resistance through deeper borrow areas preventing some areas of a pond from being flushed.

Water entering ponds is normally passed through a passive aeration unit, usually a tower (Lawson et al. 1984; Baker 1987; Hymel 1987). These are constructed of three or four tiers of successively

Figure 7. Typical temporary ditch habitat for *Procambarus clarkii* and *Procambarus zonangulus* in south Louisiana filled with rainwater and runoff. J. Huner.

smaller meshes set at 0.3-0.5 m intervals. These will normally saturate water with oxygen when properly constructed.

Many ponds are being constructed with provisions for recirculation canals (Huner and Barr 1991). That is, water enters at the highest point, flows through the sections delineated by baffle levees, and leaves through a drain at the low end. The drain empties into a canal that extends back to the entry point and is pumped back into the pond through a passive aeration unit. This reduces pumping costs substantially especially when the source of pond water is a deep well. More recently, ponds have been designed so that paddlewheel aerators used in the channel catfish industry for emergency aeration can be used to circulate and aerate the water (Hymel 1988; Bankston et al. 1989; Lawson 1990). Such ponds must be built with level bottoms if the paddlewheel units are to be effective. Continuous flow is maintained without recirculation canals, per se. It must

Figure 8. Baffle levee in a crayfish pond. This directs aerated water through the pond. J. Huner.

be noted that while such circulation systems maintain high dissolved oxygen levels, there is no conclusive proof that they ensure high, consistent crayfish production levels.

Water Management

Crayfishes tolerate low salinity brackish waters (Loyacano 1967; Perry and LaCaze 1969; Sharfstein and Chafin 1979). *Procambarus* spp. can reproduce in salinities to 3 parts per thousand (ppt) and grow in salinities to 9 ppt. Freshwaters should have total hardness and alkalinity above 50 parts per million (ppm) (de la Bretonne et al. 1969); however, excellent crayfish production has been observed in ponds with almost 0 ppm total hardness but with alkalinity above 100 ppm (Huner 1978b). In this situation, crayfish were able to obtain enough calcium from food and bottom sediments to sustain growth and reproduction. pH ranges of 6.5-9.0 are acceptable although 7.0-8.0 is preferable (Huner and Barr 1991). Cultured crayfish appear to be subjected to oxygen stress at dissolved oxygen (DO) levels below 3 ppm. At levels below 2 ppm, they will climb to the surface and use atmospheric oxygen. Oxygen is considered to be the principal water quality problem in crayfish ponds (Avault et al. 1975; Hymel 1987). Both juvenile and adult *P. clarkii* and *P. zonangulus* can persist as long as 12 hours in waters with levels of oxygen from 0.5 to 0.9 ppm at temperatures around 22°C (Huner 1987).

Thermal maxima and thermal minima are in the 34-2°C range (Huner 1987; Huner and Barr 1991). Preferred temperature for *P. clarkii* is around 22°C (Taylor 1984). No similar information is available for *P. zonangulus* but its faster growth rate in winter in Louisiana (Sheppard 1974; Romaire and Lutz 1990) suggests that it may have a comparatively lower preferred temperature. Optimal temperatures for growth of both species are not known but are certainly in excess of 21°C.

Crayfishes are sensitive to many pesticides, especially insecticides. Therefore, if surface water is used as a water source, it should be monitored for signs of pesticide contamination from nearby agricultural activities. Likewise, one must guard against aerial drift in agricultural areas. Furthermore, synergisms involving mixed pes-

Table 4. Concentrations of Selected Pesticides (mg/l) Lethal to 50% of Test Crayfish, *Procambarus clarkii,* [LC^{50}] over Short Periods, after Jarboe (1988).

Trade Name/ Common Name	Time Interval (Hours)					
Pesticide	24	36	48	72	96	120
Insecticide			LC$_{50}$	(mg/l)		
Agritox/ Trichloronat		0.81				
Aldrin/Aldrin						0.00038
Altosid/Methoprene						44.3
Ambush/Permethrin						0.00038
Anofex/DDT	0.59		0.59	0.59	0.59	
Arosurf/Isostearyl Alcohol				>10000		
Azodrin/ Monocrotophos		1.29				
BSP-1 *Bacillus sphaericus* Neide					79.2	
Bactimos/*Bacillus thuringiensis* var. *israelensis*					103.3	
Baygon/Propoxur					1.43	
Baytex/Fenthion		0.022				
Bidrin/Dicrotophos	5.55		3.75	3.02		
Bux/Metalkamate					0.28	
Cythion/Malathion					49.2	
Dibrom/Naled	5.99		4.01	4.01		
Dieldrix/Dieldrin		0.74				
Dimecron/Phosphamidon	20.0		5.99	5.52		
Dursban/Chlorpyrifos		0.041				
Dylox/Trichlorfon					0.050	
Endrin/Endrin	0.40		0.30	0.30		
Ficam/Bendiocarb					5.55	
Guthion/Azinophosethyl					0.100	
Niran E-4/Parathion		0.081				
Niran M-4/Methyl Parathion	0.057		0.04	0.04		
Scouge II/Resmethrin+ Pipernyl butoxide 1:3					0.00074	
Sevin/Carbaryl	5.0		3.0	2.0		

Trade Name/ Common Name	Time Interval (Hours)					
Herbicide	24	36	48	72	96	120
2,4, D/2,4, D					1389	
Basagran/Bentazon					701.5	
Bolero/Thiobencarb					9.2	
Gramoxone/Paraquat			5.2	2.4		
Krenite/Fosamine	2103		1684	1545	1472	
Modown/Bifenox					1338	
Ordram/Molinate					14.0	
Oust/Methyl-Sulfometuron	>10000		>10000	>10000	>50000	
Roundup/Glyphosate					41.3	
Stam/Propanil					7.9	
Treflan/Trifluralin	13.0			12.0		
Fungicide						
Aarason/Arason 70S Red					4.3	
Belate/Benomyl					1032	
Kocide/Kocide SD					2918	
Orthocide/Captan					15631	
Vitavax/Vitavax					15631	

ticides, temperatures, and life cycle stages must be considered before pesticides are applied in the vicinity of crayfish ponds or used in ricefields planted with rice during the summer (Leung et al. 1980; Ekanem et al. 1981, 1983; Naqvi et al. 1987, 1990; Gaudé 1988; Naqvi and Flagge 1990). Pesticide usage in crayfish-agronomic crop rotations should be avoided. Table 4 includes information about the relative toxicities of a number of common pesticides used in crayfish producing areas. Local cooperative extension service agents should be consulted for latest recommendations about pesticide usage in association with crayfish production.

Good quality surface water is far cheaper to obtain than well water because it seldom needs to be lifted more than 2 m (Baker 1987). It has the disadvantages of being a source of predator/competitor fishes and, in coastal areas, predaceous, euryhaline blue crabs, *Callinectes sapidus*, being of poor quality with respect to biological oxygen demand (BOD) in the autumn, contaminated with pesticides, and in short supply during the autumn when surface waters are often low in the South (Avault and Huner 1985; Huner and Barr 1991). Well water, however, must be aerated when it leaves the ground and may contain substances such as ferrous iron and hydrogen sulphide that must be eliminated before use in ponds. Ferrous iron oxidizes to ferric iron when it is exposed to oxygen. Until the oxygen demand of ferrous iron is realized, water cannot be oxygenated. The precipitate can coat and clog crayfish gills. Finally, the cost of pumping high volumes of water, especially from deep wells, can be prohibitive.

Crayfish ponds have dense stands of forage vegetation (see below) when filled in the autumn. These produce a dramatic BOD within two weeks of pond flooding (Avault et al. 1975). The only effective way to counter this adverse situation is to flush the pond with aerated water (Figures 9 and 10). Recommended flushing period is 4-5 days (Huner and Barr 1991). This management procedure requires a pumping rate of 950 liters/minute/ha for an average depth of 0.5 m. Flushing continues until BOD is reduced by lower temperatures in late autumn and is often necessary again when waters warm in the spring. Most ponds must have total volume exchanged at least 9 times each season (Baker 1987). Table 5 provides representative data for DOs as functions of time in crayfish culture systems.

Filling ponds to half final depth in the autumn reduces the pumping rate to half that needed for a fully filled pond (de la Bretonne and Romaire 1989b; Huner 1990d). Young crayfish are so small that the lower levels do not effect growth and development. The pond can be filled later by pumping water and/or catching rainwater.

One may delay planting of rice forage crops (or sorghum-sudan grass hybrid) or use a rice variety that will regrow quickly after late summer harvest to reduce BODs (de la Bretonne and Romaire 1989b; Brunson 1989a; Huner 1990d). The rice will be green and grow until the first killing frosts of late autumn come. The BOD in such systems is far lower than that in systems with much dead vegetation when they are flooded.

Some producers may cut and bale 25-50% of the vegetation in a

pond to reduce initial BOD problems (de la Bretonne and Romaire 1989; Huner 1990d). The bales are left in place in the pond and are periodically broken up to provide additional forage for the crayfish.

Crayfish ponds are normally flooded in mid-October (de la Bretonne and Romaire 1989b; Huner 1990d). This practice has generally been supported by production studies where flooding dates differed (Huner and Avault 1976b; Chien and Avault 1983; Miltner and Avault 1983). Air temperatures at that time are usually around

Figure 10. Tractor used to power a low-lift irrigation pump that pumps water from a recirculation canal through an aeration tower on the right. The tower on the left is connected directly to a well for initial flood up and water additions/exchanges, as needed. J. Huner.

25°C during the day and 20°C or lower at night. BODs are reduced at such temperatures relative to earlier flooding dates when temperatures are 5-10°C higher. Considerable pumping is required to maintain DOs at 3 ppm or higher when early floodings occur. Later floodings may subject small crayfish to increased mortality in burrows. Furthermore, growth is slowed once ponds are flooded because of lower temperatures. As a result, significant numbers of harvestable crayfish are not available before spring.

Most *Procambarus* spp. young of the year mature in mid to late spring and burrowing activity becomes apparent in levees at the water line (Huner 1978b). Crayfish ponds are generally drained at this time. Draining serves several important functions (Huner 1990d). It appears to phase crayfish reproductive activity so that most females produce young during the late summer to mid-autumn period when ponds are flooded. In addition, predator fishes are eliminated in pond systems. Ponds flooded for at least 12 consecu-

Table 5. Changes in Rice Forage Biomass, Carbon:Nitrogen Ratios, Early Morning Dissolved Oxygen, and Temperature in an Experimental Rice-Crayfish Culture System, After Day and Avault (1986).

Month	Percent Rice Remaining in Litter Bag	C:N Ratio	Early Morning Dissolve Oxygen mg/liter	Temperature °C
Oct.	—	76	—	—
24			0.4	20.3
Nov.	53.8	68	—	—
7			1.7	16.1
21			1.0	16.0
Dec.	43.5	61	—	—
5			2.3	12.5
19			2.7	13.5
Jan.	33.0	49	—	—
2			10.1	6.9
16			3.4	12.8
30			2.8	12.0
Feb.	19.8	38	—	—
13			4.4	9.9
27			2.8	16.6
Mar.	11.3	36	—	—
13			3.8	14.2
27			4.1	22.3
Apr.	3.8	30	—	—
10			6.7	17.6
24			5.6	23.3
May	3.0	22	—	—
8			2.5	22.5
22			3.8	23.7

*Initial Forage Biomass Approximately 0.2 kg/square meter.

**RE: Day, C. H. and J. W. Avault, Jr. 1986. Crayfish *Procambarus clarkii* production in ponds receiving varying amounts of soybean stubble or rice straw forage. Freshwater Crayfish 6:247-265.

tive months will normally develop dense predaceous fish populations that render them unsuitable for crayfish production. Draining
also permits the cultivation of a forage and/or marketable agronomic crop such as rice, soybeans, and sorghums. This material which
remains erect in the water column into the spring also serves as a
substrate on which crayfish distribute themselves and effectively
reduces their density per unit area. Finally, draining permits bottom
soils to oxidize preventing the possible development of deleterious
anoxic mucks, something the bottom dwelling crayfishes have difficulty avoiding.

Rice cultivation, as practiced in the USA, involves maintenance
of water levels of 10-15 cm through the summer after a short dry
period in the spring for planting. As a result, ponds are never really
dry and water tables remain high. Mature crayfish generally remain
in shallow burrows in levees during this period as the water is quite
hot. Draining for rice harvest eliminates fish, if present. It is not
clear if such high water levels/tables have any effect on crayfish but
the crayfish are clearly less subject to drought stress from low water
tables. This is not the case with crayfish in ponds that are completely drained.

It should be emphasized that there have been no studies to determine the effects of different pond draining dates on crayfish production. Research ponds have been flooded for 12 or more consecutive months in Louisiana and in Mississippi and have generated
crops in excess of 1000 kg/ha/yr where fish predators were not a
problem and provisions were made to provide supplemental forages
and/or feeds (Huner et al. 1983a; Niquette and D'Abramo 1989,
1991).

Forages and Feeds

Crayfishes are polytrophic (Lorman and Magnuson 1978; Momot et al. 1978) consuming all manner of living plants and animals
as well as microbially enriched detritus, principally plant detritus.
In the absence of predators, they will eliminate palatable vascular
vegetation (Lodge et al. 1985; Feminella and Resh 1989). Crayfishes can also damage fish in nets when abundant (Lowery and
Mendes 1977).

The bulk of crayfish carbon is derived from the living epiphyton on the surface of the plant detritus that they consume (Sanguanruang 1988). However, essential nutrients including amino acids and fatty acids as well as carotenoids and other critical micronutrients are derived from eucaryotic animals and plants especially zoo- and phyto-plankters, respectively. Crayfishes simply do not grow well in the absence of animal protein in their diets. Crayfishes have a very efficient filter-feeding apparatus composed of setae on the oral appendages (Budd et al. 1978).

A most interesting assemblage of invertebrate animals, most of which are consumed by crayfish and some of which consume crayfish, develops in crayfish ponds after they are filled (Barr et al. 1978; Huner and Naqvi 1984). Excellent reviews of crayfish food requirements and natural foods include those of Kossakowski (1966), Goddard (1988), D'Abramo and Robinson (1989). Reports on food habits in crayfish ponds include those of Huner and Naqvi (1984) and Sanguanruang (1988).

A major consideration of those studying detrital food chains such as those established in crayfish ponds is C:N ratio. This ratio falls as vegetation is colonized by epiphyton. The rate of decrease is determined by the nature of the vegetation and is most rapid where levels of ligins and cellulose are lowest. The so-called ideal C:N ratio is 17:1 (Goyert and Avault 1977). Representative C:N ratios for various fodders over time (Avault et al. 1983; Day and Avault 1986) are presented in Tables 5 and 6. This will serve as a guide for those considering use of various materials as supplemental fodders for crayfish. One vital function of vegetation however is to provide substrate. Therefore, use of rapidly decomposing vegetation may not be the best practice for extensive type crayfish management as it is now practiced.

Formulated feeds are not widely used in pond crayfish culture. Most crayfish are raised in ponds in which a vegetative forage crop is established during the summer dry period (Brunson 1989a; de la Bretonne and Romaire 1989b; Avault and Brunson 1990; Huner 1990d). This crop, then, supplies the detrital base for a natural food web that has occasionally generated crayfish yields approaching 3000 kg/ha/yr, although average production is much lower at 500-600 kg/ha/yr (Table 3). Target forage production is 0.5-1.0 kg

Table 6. Variation in Carbon:Nitrogen Ratios of Various Vegetative Forage Materials, for Crayfish, *Procambarus* spp., Under Aerobic Conditions at Approximately 22°C, After Avault et al. (1983).

Forage	Initial % Protein	C:N 0	Ratio 2	(weeks) 6	12
Soybean stubble	18.8	13	11	10	7
Soybean	44.8	6	7	6	7
Sweet Potato Vine	11.8	20	17	11	10
Sweet Potato Trimmings	7.1	25	14	6	6
Rice Stubble	5.5	45	38	22	13
Annual Grasses	10.4	19	18	11	10
Rye Hay	9.3	18	18	10	11
Corn Leaves	17.1	15	12	8	8
Sugar Cane Bagasse	3.6	57	89	36	38
Mixed Deciduous Tree Leaves	6.4	61	35	27	28

*Sweet Potato – *Ipomea balatas;* Rice – *Oryza sativa;* Rye – *Secale cerale;* Soybean – *Glycine max;* and Sugarcane – *Saccharum officinarum.*

**RE: Avault, J. W., Jr., R. P. Romaire, and M. R. Miltner. 1983. Feeds and forages for red swamp crawfish, *Procambarus clarkii:* 15 years research at Louisiana State University reviewed. Freshwater Crayfish 5:362-369.

dry matter per square meter. Higher levels have been shown to create unmanageable BODs (Morrissy 1979). Biomasses generated by representative forage crops are presented in Table 7.

Fodder management depends on the nature of the pond system in use. However, the recommended forage in permanent open ponds is some variety of rice (Table 8) that resists lodging (collapsing), is long lived (low senescence rate), and produces a high biomass per unit area (Brunson 1989a, c; Avault and Brunson 1990). Millets are generally not recommended for crayfish pond forage because they die soon after being flooded and decompose very rapidly. Annual and perennial volunteer vegetation has been and continues to be used by some crayfish farmers but crayfish production lags significantly behind that realized when rice is used as forage (Chien and Avault 1979; Avault et al. 1983; Brunson 1989a; Avault and Brunson 1990).

Table 7. Forage Biomass for Crayfish, *Procambarus* spp., Generated by Various Forage Plants, After Brunson (1989c).

Forage	Culture System/ Management Methods	Dry Matter kg/ha
Rice	Double Cropped	5,000
Rice	Monoculture	2,000
Grain Sorghum (Milo)	Double Cropped	4,000
Grain Sorghum (Milo)	Monoculture	2,000
Sorghum-Sudangrass Hybrid	Early Summer Planting	8-10,000
Sorghum-Sudangrass Hybrid	Late Summer Planting	2,000
Japanese Millet	Mid-Summer Planting	3,400
Native Plants	Volunteer - not planted	2,900
Browntop Millet	Mid-Summer Planting	2,500

Annual plant forage, intentionally planted or volunteer, disappears as a consequence of the foraging activities of crayfish and natural mineralization. Witzig et al. (1983) presented two regression equations to predict loss of forage in a crayfish pond planted with rice as a forage base. Overall dry biomass was 676 g per square meter but some areas had higher levels than others so they divided the pond into areas containing either greater than or less than 647 g per square meter. Vegetation disappeared after about 220 days at the low density and 180 days at the higher density. Biomass reached its highest point 4 weeks after the pond was flooded as the rice continued to grow actively until then. The two equations were:

Less than 647 g per square meter
$y = 0.515 - 0.00233x$; R Square = 0.82

Greater than 647 g per square meter
$y = 1.219 - 0.00624x$; R Square = 0.88

where x = number of days after flooding and y = remaining vegetation in g per square meter.

Stunting of crayfish is the invariable result of forage depletion

Table 8. Rice Varieties Recommended for Forage in Crayfish Ponds, After Brunson (1989a).

	Culture System	
Grain Type	Double Crop	Monoculture
Long	Labelle	Starbonnet
	Lemont	Newbonnet
		Bellevue
Medium	Mars	Mars
Short	Nortai	Nortai

especially in ponds with very poor forage crops. Farmers have tried to counter this by adding hay prior to the time that the forage disappears (Figure 11). While data generally show that this will improve yield (Huner and Romaire 1979; Romaire et al. 1979; Day and Avault 1986), it is not an entirely satisfactory substitute for a dense, in-pond forage crop. Witzig et al. (1983) presented the following regression equation to show the relationship between days following pond flooding and stunting based on the pond described immediately above. This is fairly characteristic of a "good" pond as production was around 1200 kg/ha. The equation is:

$$y = -83.39 + 1.40\,x - 0.0042\,x^2 \quad R \text{ square} = 0.65$$

where y is average weight in grams and x is days following flooding.

Agronomic rotations invariably involve the following scenarios: rice-crawfish-rice and repeat; rice-crawfish-soybean-fallow and repeat; and, more recently, rice-crawfish-grain sorghum-crawfish and repeat (Brunson 1989a, b, c). Details of planting dates and recommended varieties in Louisiana are presented in Table 9. In the first rotation, crayfish are stocked initially and do not normally have to be restocked. However, continuous flooding of rice fields for rice and crayfish production reduces rice yields so soybeans, a dry soil crop, have been added to the rotation cycle. Soybeans are normally poor producers of vegetative biomass so fields are usually not flooded during the autumn following soybean harvest. Rather, the rotation begins again with rice in the spring and brood crayfish being stocked again. Substitution of grain sorghum for soybeans chainwhere economic considerations are favorable permits autumn flooding for crayfish following the dry ground crop because the sorghum produces a substantial forage biomass. The cycle can then be repeated with rice planted in the following spring and no restocking of crayfish is necessary.

Recent work with sorghum-sudan grass hybrid has suggested that it might be useful as a forage (Brunson 1989b). It is a crop raised for general livestock grazing, hay, and silage. However, if it is planted too early and not cut periodically, central stems become too thick and too tall to use as an effective crayfish pond forage. It does not grow well in water but is very tolerant of flooding so it does not die quickly when it is flooded as do millets.

Perennial vegetation in crayfish ponds most often mentioned in the literature includes alligator weed, *Alternanthera philloxeroides*, smartweeds, *Polygonum* spp., and water primroses, *Ludvigia* spp.

Figure 11. When there is too little forage for crayfish, hay may be added to the pond. Bales and rolls of hay must be broken up so that the relatively small crayfish can access it. J. Huner.

Table 9. Crayfish and Agronomic Crop Rotations, After Brunson (1989) and de la Bretonne and Romaire (1989b).

A. Crawfish Monoculture

Month(s)	Management Action(s)
April-May	Stock 50-60 kg of adult crawfish/ha.
May-June	Drain pond over 2-4 week period.
June-August	Plant suitable forage crop. Rice is recommended. Sorghum-Sudan Grass hybrid may be used in some cases especially during very dry periods.
October	Reflood pond.
November-May/June	Drain pond and repeat the cycle. Restocking is normally not necessary.

B. Rice – Crawfish – Rice

Month(s)	Management Action(s)
March-April	Plant rice.
June	At permanent flood, rice 20-25 cm high, stock 50-60 kg of adult crawfish/ha.
August	Drain pond and harvest rice.
October	Reflood pond.
November-April	Harvest crawfish.
April	Replant rice. Restocking of crawfish is usually not necessary.

C. Rice – Crawfish – Soybeans (or Grain Sorghum, Milo)

Month(s)	Management Action(s)
March-April	Plant rice.
June	At permanent flood, rice 20-25 cm high, stock 50-60 kg of adult crawfish/ha.
August	Drain field and harvest rice.
October	Reflood pond.
November-May	Harvest crawfish.
Late May-June	Plant soybeans.
October-November	Harvest crawfish.

Options:

– Reflood pond and harvest crawfish. Forage base is very poor.

– Do not reflood pond but leave fallow until spring and repeat rice cycle and restock craw fish.

– Substitute grain sorghum, milo, for soybeans and reflood the pond in October. Forage biomass should be sufficient for good crawfish crop. Repeat rice cycle and do not restock crawfish.

(Brunson 1989a; Avault and Brunson 1990; Huner 1990d; Huner and Barr 1991). These provide excellent substrate for crayfish and many food items and, in the cases of alligator weed and water primroses, excellent sources of carotenoid pigments. These plants are weeds in ricefields and can grow so rapidly that they interfere with water flow and harvesting activities in permanent crayfish ponds. Ponds dominated with semi-aquatic perennial vegetation are generally never as productive as ponds planted with rice.

Johnson et al. (1983) discuss a comparison of the semi-aquatic delta duck potato, *Sagittaria graminea platyphylla*, with rice as a forage for crayfish. Yields were similar, over 1000 kg/ha, but more crayfish were produced in autumn in the duck potato treatment, apparently because there was a significant amount of epiphyton for the crayfish to eat and early morning oxygen levels never descended to stressful levels. Two significant cohorts of harvestable crayfish, autumn and spring, developed in the duck potato ponds but only one cohort, spring, developed in the rice ponds. Delta duck potato is very attractive to waterfowl and might be useful where farmers wish to combine waterfowl hunting with crayfish production (see below).

It has long been known that crayfish will eat formulated fish feeds and grow well (Smitherman et al. 1967; de la Bretonne et al. 1969; Clark et al. 1975). Costs and returns involved with using feeds were not in balance; however, so there was little real interest in using them. The recent emphasis on graded crayfish (see below) has changed this situation (Huner 1990d).

Formulated crayfish feeds have been developed using data from basic nutrition studies (Huner and Meyers 1979; Brown et al. 1986; Davis and Robinson 1986; Hubbard et al. 1986; Reigh et al. 1990). D'Abramo and Robinson (1989) make the following general recommendations for a practical crayfish feed: protein (2:1 plant/animal), 30-35%; lipid, 3-6%; carbohydrate, 20-30%; fiber, 10%; protein/energy ration, 80-120 mg protein/kcal; and energy level, 2.5-3.0 kcal/g. Brown et al. (1986) and Reigh et al. (1990) reported that *P. clarkii* are able to more efficiently use grain byproducts than high fiber roughage or animal protein products based on dry matter, crude protein, and energy digestibility coefficients.

Two years of pond feeding studies using commercial 25% protein crayfish feeds have yet to demonstrate that use of such feeds is cost effective (Huner, unpublished; Romaire 1989a; R. P. Romaire, School of Forestry, Wildlife, and Fisheries, Louisiana State University, Baton Rouge, Louisiana, USA, personal communication). Growth responses have been demonstrated but size and/or production have generally not been great enough to justify the added expense of feeding the crayfish. Apparently, the interaction between density and timing of the initiation of feeding needs to be clarified before producers can effectively determine potential costs and returns from using crayfish feeds.

Interestingly, range cubes, compressed grain dust cubes used for cattle feed, with 9-10% protein levels have doubled crayfish production in experimental ponds when used as supplements to rice forage (Cange et al. 1981). Manufactured crayfish baits with 17-18% protein and agronomic crops including soybeans, rice, corn, and sorghum have generated good growth in crayfish grown in cages in ponds (Huner, unpublished data).

Farmers have, in fact, been providing supplemental "formulated feeds" to their crayfish for years (Huner 1990d). That is, the formulated baits and fresh/frozen fish used for crayfish bait have been consumed by small crayfish that entered and left traps at will as well as by larger crayfish that found their way out of traps. Bait usage is about 6.3 kg/ha/day when 50 traps/ha are baited with 125 g of bait per trap. Depending on size and population density of the resident crayfish population, as little as 5% or as much as 80-90% is exported back into the pond each day.

Population Management

The recommended *Procambarus* spp. for pond culture is *P. clarkii*. Consumers prefer this species over *P. zonangulus* in the Louisiana industry which dominates all North American crayfish industries (Huner 1989d). White river crayfish are often present where ponds are built, but they have not normally done well with *P. clarkii*. They usually account for less than 20% of the harvest in Louisiana crayfish ponds (de la Bretonne and Romaire 1989b; Huner 1990d; Romaire and Lutz 1990). However, the percentage of *P. zonangulus* can be much higher in local situations, a matter discussed later.

The recommended stocking rates vary according to the presence/ absence of native commercial crayfishes in the pond area (de la Bretonne and Romaire 1989b; Huner 1990d). Sites with good cover to protect burrowing crayfishes and native crayfishes should be stocked with around 25 kg/ha while those without these characteristics should be stocked with 50 + kg/ha. Stocking is normally done in mid to late spring when mature crayfish are abundant. Crayfish that will be stocked should not be refrigerated (Huner 1990d). They should be kept out of water a minimum of time and stocked during a cool period of the day.

The majority of stocked crayfish should be mature, Form I, sex ratio should be 1:1, and at least 20% of the females should have well developed, brown colored eggs in their ovaries (de la Bretonne and Romaire 1989b). Crayfish for stocking should come from a nearby source, be protected from dehydration and damage during transport, and be stocked as soon as possible after capture. Heat and cold shocks should be avoided. Ponds may be a better source of broodstock because wild crayfish seem to be more apt to leave a pond, especially those caught in moving water systems like the Atchafalaya Basin (Huner and Barr 1991).

No real effort has been made to correlate stocking density with production. Production has two components: the number of crayfish caught and/or the volume harvested. Obviously, the greater the number harvested, the greater the total volume. However, the inter-action between population density and growth is such that average size in one system may be smaller compared to that in another but the same volume by weight comes from each system (Lutz and Wolters 1986). Literature on the relationship between stocking rates and average size of the harvested crayfish is lacking although Ever-sole (1990) alluded to reduced size in one series of experimental ponds as a result of higher densities, and presumably greater numbers of young in the following year. Two studies report no difference in production as volume by weight of *P. clarkii* at stocking rates from 0-100 kg/ha but do not provide data on average size which might well show that more but smaller crayfish were harvested from the ponds with the higher stocking rates (Chien and Avault 1980; Avault and Granados 1990). It is clear that all ponds had pre-existing populations to start!

It has become obvious that most of the burrows in crawfish ponds are in the levees. Few crayfish are able to successfully burrow in the anoxic muds on pond bottoms when ponds are drained. As a result, ponds with extensive levee area per unit area, either small ponds or large ponds with many baffle levees, will have greater production potential than ponds without as much levee/area. This complicates management and further emphasizes the problem of projecting production from one season to the next.

The significance of pre-existing crayfish populations in a pond site has not been properly emphasized in management texts. South Carolina is a state without significant numbers of native commercial crayfishes. As a result, its crayfish industry was initiated by importing *P. clarkii* and, presumably, some *P. zonangulus,* from Louisiana. Stocking rates of approximately 50 kg/ha were used. First year production in well constructed, properly managed ponds has been about 500 kg/ha (Eversole and Pomeroy 1989; Pomeroy et al. 1989). Second year production increased by factors of 1.5-2.5 in many cases. These results can be explained by low survival of initial broodstock with poor production of young and subsequent increased survival of second generation broodstock. This pattern can be seen in Louisiana where there was no pre-existing broodstock in the area being used for a crayfish pond. However, a number of sites have pre-existing populations and first year production of 1000 kg/ha or more is often seen in such situations.

All *P. clarkii* do not spawn simultaneously and gravid females are found in any month (Penn 1943; Huner 1975, 1978b; Huner and Avault 1976b). However, the majority seem to produce their young from late summer through mid-autumn. These young are released once the pond is flooded or while it is flooded. Several cohorts appear at intervals through the autumn into early winter. In contrast, *P. zonangulus* seem to release young over a more restricted period in mid-autumn (Sheppard 1974; Huner 1990d; Romaire and Lutz 1990).

Large numbers of young and optimal, but as yet undefined, conditions are required to generate yield over 1000 kg/ha. Loss of young, poor forage crops, or indifferent harvesting effort especially when populations are dense insure a poor crayfish crop made up of either many, relatively small, low value crayfish or few, but high value, large crayfish, irrespective of species (Huner and Barr 1991).

Some small *P. clarkii* appear in late winter to early spring in Louisiana crayfish ponds (Huner 1978b; Romaire and Lutz 1990). These seem to play a very reduced role in production in that season. However, they may constitute a significant number of medium-sized holdover crayfish that survive in burrows until the next season begins.

The reproductive potential of a normal pond-reared female *P. clarkii* (9.0 cm total length) is 250-300 young (Penn 1943; Suko 1958). Therefore, with 100% survival in a new pond stocked at 25 kg/ha of 9 cm (20 g) crayfish, sex ratio of 1:1, one would expect a minimum of 156,250 young and potential production of 3125 kg/ha. With industry wide production of 500-600 kg/ha/yr, it is easy to see that survival (and/or growth) of brood crayfish and/or young crayfish is very poor relative to potential production.

Predicting production is more of an art than a science. Young crayfish may be caught with small mesh dip nets and/or heavily weighted, small mesh seines. Heavy vegetation precludes use of anything but heavy duty sampling devices for much of a crayfish season. Momot and Romaire (1981) and Huner and Barr (1991) describe methods to project production on the basis of relative numbers of young in dip net and seine catches and growth rates over sequential 1-2 week sampling intervals.

When small crayfish are easy to catch, potential production is great but mean size is likely to be small. If the crayfish are not intensively harvested when abundant, however, growth of young cohorts may be stifled to the extent that they never grow large enough to harvest and production is poor. When crayfish are abundant, harvesting may be profitable from mid-autumn through mid to late spring with periods of poor catch during the coldest period of winter. If small crayfish are not abundant, mean size will be substantially larger but harvesting will probably not be cost effective until mid-spring, if at all.

Until recently farmers have had only one option when crayfish are scarce in ponds. That is, they could harvest intensively in mid to late spring when the main cohort matured. Until that time their catches normally would be very poor, not more than a few crayfish per trap. Overall production would be a few hundred kg/ha although crayfish would normally be very large. Restocking with more adult crayfish has been the only prudent recommendation that advisory agencies could make in this situation.

A new management option is now available and arose because high prices are now being paid for very large crayfish and very low prices are being paid for small crayfish, if they can be sold at all. As a result, the transfer of small 5-10 g immature crayfish from high density ponds to low density ponds at rates of 100-300 kg/ha in late winter to early spring has, in some cases, provided an economically viable alternative to the preceding option (de la Bretonne and Romaire 1989a; Huner 1990d; 1992). Mean size may exceed 35 g in transferred crayfish depending on survival and growing conditions. The same crayfish may not reach 15 g in the source pond.

McClain and Bollich (1992) have documented an effective transfer technique called "relaying." Here small, immature *P. clarkii* are stocked in newly planted rice fields in May. The crayfish are then harvested in July, long after most regular crayfish ponds are drained. Stocking rates were 500 and 1000 kg per ha with equivalent recovery. The crayfish increase in size 2-4 times.

Crayfish in high density ponds are small and are often in poor physiological condition (Huner and Romaire 1979; Huner and Barr 1991). Even those females that are in "good" condition are less fecund than females from populations that are not density limited because they are smaller. Those in poor physiological condition might not survive the summer and, if they do, fecundity may be reduced.

Long-time Louisiana crayfish farmers who cultivate crayfish in a number of different but often contiguous ponds report a cycle that suggests serious problems resulting from the inability to control population densities. That is, production per unit area will be poor in the first year, less than 500 kg/ha, after a new pond is stocked but size will be large. Production per unit area will be very good, 1000-2000 kg/ha, in the second year but size will be moderate. Production per unit area will be poor, less than 500 kg/ha, in the third year because crayfish, while abundant will be small, with many being too small to justify harvest. Finally, crayfish will be virtually absent but large in the fourth year. Overall production per unit area over the entire farm will be at the state average of 500-600 kg/ha because different ponds will be at different stages (1990, J. Boyce, Boyce Crawfish Farm, Sorrento, Louisiana, USA and L. Richard, Richard Farm, Lawtell, Louisiana, USA, personal communications).

There are few data on the interactions of *P. clarkii* and *P. zonangulus.* Lutz and Wolters (1987) grew the two species together or alone in pools planted with rice for forage. Density of 25-35 mm young was 6 per square meter with equal numbers when the two species were stocked together. Growth was similar in all treatments but survival was 15-20% greater for *P. clarkii.* Huner and Pfister (1990) observed similar results in a 0.4 ha earthen pond stocked with equal numbers of 25-50 mm *P. clarkii* and *P. zonangulus* at 2.5 per square meter. However, in a second pond, they stocked the two species at a ratio of 9 *P. clarkii* to 1 *P. zonangulus* at a rate of 5 per square meter. Survival of *P. zonangulus* was negligible.

Romaire and Lutz (1990) studied population dynamics of *P. clarkii* and *P. zonangulus* in two large commercial crayfish ponds. Survival and growth were similar for the two species. *Procambarus clarkii* was the dominant species, 7-14 to 1. Final density of all harvested crayfish was about 3 per square meter. These data, then, contradict the results of the two stocking studies where *P. clarkii* negatively impacted survival of *P. zonangulus.*

Evidence is conclusive that the white river crayfish is an annual spawner, spawns later in the autumn, grows faster at lower temperatures, grows larger, and is absent from highly eutrophic, poor water quality habitats such as back swamps and the marsh where *P. clarkii* are abundant (Penn 1956; Sheppard 1974; Huner 1975; Hobbs and Jass 1988; Romaire and Lutz 1990; Huner and Barr 1991). These factors appear to give the white river crayfish a temporal advantage over *P. clarkii* in conventional ponds where *P. clarkii* are scarce before white river crayfish recruit. However, there appears to be no way, at this time, to project which species will be abundant from one year to the next although ponds with large numbers of white river crayfishes seem to be more likely to have them in succeeding years. This situation has changed from earlier observations that *P. clarkii* normally became dominant in most ponds (Huner and Barr 1991). Note: Evidence (Huner, unpublished) is mounting that

P. zonangulus is resistant to parasites and pathogens that are lethal to *P. clarkii*. Perhaps toxicants have been applied too late to kill many of the *P. zonangulus* which, presumably, have already burrowed out of harm's way.

Attempts to eradicate *P. zonangulus* populations in Louisiana crayfish ponds with toxins in the spring followed by restocking of *P. clarkii* have not succeeded in shifting the population balance. An interesting observation about *P. zonangulus* is that they are more abundant in trap catches during the late winter through mid-spring regardless of their absolute numbers. They also mature earlier (Huner and Paret 1990).

Research and extension efforts have emphasized improvement of water quality, especially dissolved oxygen, through development of efficient pond aeration systems. Modified ponds require numerous baffle levees to channel oxygenated water uniformly. The levees increase the sites available for burrowing. The improved water quality and increased area of baffle levees no doubt, impact production and species composition but this has not been investigated as yet.

Summer production of *P. clarkii* in the South is possible (Table 10) (Huner et al. 1983a, b; Culley et al. 1986; de la Bretonne and Romaire 1989a). Young of the year that appear in mid to late spring will grow if ponds are left flooded into the summer. Attempts to produce such crayfish have generally involved draining the ponds during the winter or early spring and reflooding in mid to late spring. Some forage crop may be grown during this dry period including wheat or rye, if planted in the winter, or rice, if planted in early spring. An alternative to this management practice is the stocking of undersized, immature crayfish from ponds with potential stunting problems.

Water temperature in the typical, shallow water crayfish pond becomes far too high in the summer months for *P. clarkii*. Where this species has been produced commercially in the summer, farmers have planted rice varieties that generate a canopy to insulate water from solar radiation. Furthermore, water depth was raised to 15-30 cm compared to the normal 10-15 cm. Elsewhere, work has been done in relatively deep (1.0-1.5 m) experimental ponds at universities where water depth provided thermal insulation for the

Table 10. Off-Cycle Crayfish Management Methods, After de la Bretonne and Romaire (1989a).

Louisiana Monoculture Cycle

Month(s)	Management Action(s)
November-December	Stock brood crawfish; drain pond; plant forage.
March-April	Plant and grow forage such as oats, rye grass, wheat, etc.
April-May	Fill ponds with water 0.6-1.0 m. Natural recruitment of young crawfish.
June-October	Harvest crawfish.

South Carolina Off-Cycle Crawfish Management Scheme

Month(s)	Management Action(s)
February	Use established pond. Fill pond with water to a depth of 1 meter. Natural recruitment of young crawfish.
February-September	Feed supplemental hay and alligatorweed.
June-September	Harvest crawfish with traps.
October-November	Drain pond and repeat cycle.

Texas Off-Cycle Crawfish Management Scheme

Month(s)	Management Action(s)
March	Plant rice as crawfish forage.
March-May	Raise water level as rice grows.
May-June	Supplementally stock small, immature crawfish (7-10 g) at 400 to 500 kg/ha in 0.5 m of water.
July-September	Harvest crawfish with traps.
October-March	Drain pond or keep pond filled for fall, winter, and spring production.

crawfish (Huner et al. 1983a,b; Culley et al. 1986; Niquette and D'Abramo 1989; D'Abramo and Niquette 1991).

Reproducing populations of *Procambarus clarkii* are present in the following states: Delaware, Maryland, Virginia, North Carolina, South Carolina, Georgia, Florida, Alabama, Mississippi, Louisiana, Texas, New Mexico, Arizona, Nevada, California, Oregon, Utah, Idaho, Arkansas, Missouri, Tennessee, Kentucky, Illinois, Indiana, and Ohio (Hobbs 1972; Huner and Barr 1991; Hobbs et al. 1989). It is present in ricefields in the Sacramento-San Joaquin Delta in central California where Sommer and Goldman (1983) and Sommer (1984) have studied its life cycle and suggest that it has cultural potential there. Huner (1984) discusses culture of *P. clarkii* and white river crayfish in other areas besides the deep South. This practice involves shifting away from an autumn-spring to a spring-autumn culture program because growth cannot take place during colder periods of the year. These become progressively longer as one moves northward.

Romaire (1990) has developed a computer simulation of commercial crayfish culture ponds based on *P. clarkii*. This simulation has proven especially useful in projecting effects of harvesting strategies on population structures and harvest. It is based on field and laboratory data and is currently being verified with real time data. Computer driven operation is a tool long needed by the crayfish industry. The simulation's value will increase as information about white river crayfish is factored into it and additional field data on *P. clarkii* are generated.

In northern latitudes, young *P. clarkii,* white river crayfishes, and *O. immunis* are usually not released until spring. Ponds should be maintained so that crayfish do not die in burrows. This is usually accomplished by keeping water in them over winter (Huner 1976). If forage vegetation is grown, it is usually a hardy plant like rye.

Harvesting

Harvesting is a critical consideration in crayfish culture. Costs of bait and labor constitute 60-80% of the annual operating costs (Dellenbarger et al. 1987; Pomeroy et al. 1989; Romaire 1989b) (see Economics section). Therefore, much research has been directed in this area.

Crayfish are harvested almost exclusively with baited mesh traps (Romaire and Pfister 1983a; Romaire 1988, 1989b; Huner 1990d; Huner and Barr 1991). The general design is a vertical cylinder, open at the top, with two or three funnel-shaped entrances at the base. Plastic coated hexagonal poultry netting with a 1.9 cm mesh is the conventional trap building material. A vertical rod serves to hold the trap upright. Rods may be attached to traps or placed in the pond with traps suspended on them. An impassable retainer ring prevents crayfish from climbing out of the open top. Height can range from 0.67-1.25 m depending on water depth. The three basic designs now in common use are described by each trap's physical appearance. These are the pyramid trap (Figure 12), the barrel trap, and the pillow trap (Figure 13). Funnels are made in pillow and pyramid traps by inverting a 5.0-6.5 cm opening at a corner with a broad-shouldered bottle. Funnels are constructed separately and sewn into the sides of barrel taps. Pyramid traps are currently considered to be the most effective traps.

Traps are set so that they rise above the surface to facilitate checking and to provide crayfish with access to the surface if oxygen declines to stressful levels (Huner and Barr 1991) (Figure 14). Floats are frequently attached to traps to facilitate recovery when they are knocked down (Figure 15). Trap densities have ranged from 50-75 traps per ha although new data (Romaire and Huner, unpublished) suggest that densities of 25-50 per ha may be more cost effective if a management priority is placed on growing larger crayfish.

Traps are typically checked once a day (Pfister and Romaire 1983a; Romaire and Pfister 1983b). When crayfish are especially abundant, it may be prudent to bait and check traps twice each day; however, 12 hour sets invariably produce smaller crayfish than do 24 hour sets. Escape rate is high when sets exceed 24 hours. Traps have been traditionally checked and baited in the morning but catch of large crayfish is usually increased if traps are checked in the morning with rebaiting in the late afternoon (Huner et al. 1989). Larger crayfish are more active at night and bait may be consumed and/or lose its effectiveness when put into traps in the morning.

Farmers use common sense in planning the trapping effort. If market conditions are poor, effort is low. This approach can create problems in ponds with high population densities because failure to remove the largest crayfish, regardless of their absolute density, inhibits growth of the smaller crayfish (Momot and Romaire 1981). Catch per unit effort is affected principally by density and temperature (Romaire and Osorio 1989) and it is highest when both of these factors are high, usually in late season. Moon phase, molt cycle, and weather changes have a reduced, but identifiable, impact on catch.

In general, farmers rarely check traps in any pond more than 3 consecutive days in a week unless catch and/or prices are high enough to justify the effort. Farmers will typically rebait traps that have been unbaited without removing the crayfish. New data (Huner and Romaire, unpublished) show that, depending on the pond, significant quantities of crayfish will enter empty traps and should be harvested as traps are rebaited.

Bait was traditionally some form of low-value frozen fish of the

Figure 12. Pyramid crayfish trap. Note the three funnel entrances at the corners of the base. The support rod is obscured by a trap seam. The plastic retainer ring also serves as a handle. J. Huner.

Figure 13. Verticle pillow trap. Note the two funnel entrances at either corner on the base, the support rod with a handle, and the open left corner with a retainer ring made of aluminum flashing. J. Huner.

families Clupeidae–gizzard shad, *Dorosoma cepedianum*, skipjack herring, *Alosa crysoleucas*, and menhaden, *Brevoortia* spp., Catastomidae–buffalofishes, *Ictiobus* spp., and suckers, *Catostomus* spp., and Cyprinidae–common carp, *Cyprinus carpio* (Huner et al. 1989; Romaire 1989b; Huner 1990d). In times of bait shortage, fish offal of many species as well as by catch from the shrimp trawling industry have been used as crayfish baits. Pieces of nutria (*Myocas-*

Figure 14. Barrel traps in pond. Note that the tops extend above the surface and that the entire top of each trap is open but each has a retainer ring made of aluminum flashing. J. Huner.

tor coypu) carcasses have also been used during cold months when these aquatic rodents are trapped for their fur (Huner and Barr 1991).

Fish, although effective as crayfish bait, has the following disadvantages: freezers are required to keep large volumes; cutting bait is unpleasant and creates health hazards; and discarding of old bait creates sanitary problems (Romaire 1989b; Huner 1990d). In addition, the dramatic increase in crayfish culture in past years created an equivalent demand for bait which could not be easily met by the commercial fishing sector so prices rose. All these factors have led to the development of artificial, or manufactured, crayfish baits.

Manufactured crayfish baits are compressed pellets of ground

Figure 15. Checking a barrel trap in a crayfish pond. Note the ring of flotation material at the top. Flotation rings and floats are used to facilitate recovery of fallen traps. J. Huner.

grain/seed products such as corn meal, rice bran, soybean meal, cottonseed meal, and wheat flour, grain dust recovered from grain storage systems, fish meal, proprietary attractants, and binders such as inert bentonite, a proprietary plastic polymer or organic sorghum (milo) flour (Burns and Avault 1985; Cange et al. 1985-86; Cange et al. 1986b; Romaire 1989a; Huner 1990d). Pellets are round and have a diameter of 5.0-7.5 cm. Protein levels may range from 15-25% although one bait, no longer on the market, had a level around 60% (Huner et al. 1991). Water stability may be as little as 3 or 4 hours or indefinite depending on the quality and quantity of the binder.

Research has shown that 110-125 g of bait per trap is the most cost effective quantity per set (Huner et al. 1989; Romaire and Osorio 1986). Furthermore, at water temperatures below 20-25°C, a combination of 50% cut fish and 50% manufactured bait is more effective than either used separately (Huner et al. 1989, 1990b; Burns and Avault 1990; Huner and Paret 1990). Above 22°C, manufactured baits are as effective or more effective than the combination. Available studies have not shown clear-cut differences between types of fish used for crayfish bait but softer fish are consumed more quickly than tougher fish with a reduction in catch because crayfish leave traps and are not replaced. Some manufactured baits are superior to others in catch per unit effort but may not be cost effective as a cheaper but less attractive bait (Romaire and Osorio 1986).

While men walking with a boat or push-poling a boat still harvest crayfish commercially, most crayfish are harvested from motor driven boats (Romaire 1989b; Huner 1990d). Boats are relatively wide, flat-bottomed units with square ends being about 2 m wide × 5 m long. The most common harvesting systems are the Go Devil and the Crawfish Combine.

The Go Devil system utilizes a boat powered with a stern-mounted, air-cooled engine that has a long, straight shaft and weedless propeller (Figures 16 and 17). Normally, two men operate the unit (Romaire 1989b; Huner 1990d). One man in the bow lifts traps, empties them into a sorting table and resets them. A man in the stern rebaits the traps and steers the boat. A modification of this system involves the use of a cleated wheel to push the boat rather than a propeller. Another modification involves a reconfiguration of the steering mechanism from a tiller to a foot-controlled steering bar permitting one man to operate it and check traps at the same time.

The Crawfish Combine has a bow- or stern-mounted cleated wheel which normally is powered by hydraulic fluid pumped by an air-cooled engine (Romaire 1989b; Huner 1990d) (Figures 18 and

Figure 16. A one-operator "Go Devil" powered harvest boat. J. Huner.

19). The wheel pushes or pulls the boat. The operator controls direction with foot pedal controls and may also check traps by himself resulting in a significant reduction in labor.

An individual walking or push-poling a boat in a pond can check 40-50 traps per hour. Go Devil or Crawfish Combine systems can be used to check over 200 traps per hour (Romaire 1989b; Huner and Barr 1991).

Seine and trawl systems are not widely used to harvest crayfish for two reasons (Huner 1990a, d). First, they capture all crayfish in their paths so that sensitive and fragile premolt, soft, and recently molted postmolt crayfish are caught as well as the stable intermolt specimen attracted to traps. The weak crayfish must be sorted from the strong ones, an additional expense warranted only if the sorted crayfish have high enough value to justify the added expense. Second, seines and trawls are difficult to use in heavily vegetated crayfish ponds.

D'Abramo and Niquette (1991) reported good catches of *Pro-*

Figure 17. "Go Devil" engines showing tillers and weedless propellors at the end of long, straight shafts. J. Huner.

cambarus spp. in permanently flooded, debris-free ponds with a hand-pulled seine. While overall production was one half to two thirds of that in control ponds harvested with traps, the batch harvest required no bait and minimal labor.

Entrepreneurs and researchers have attempted to develop automated crayfish harvesting systems involving both trawls and traps (Huner 1989b, 1990a, d). In the case of trawls, beam trawls with fixed, rigid openings have been the preferred trawl configuration. These have been constructed from some form of aluminum frame with tubing or plate or from a steel frame with a polyethylene trawl net webbing. Several systems have limited commercial application

Figure 18. A Crawfish Combine with a bow-mounted, hydraulic propulsion wheel. J. Huner.

and are patented.[2] One utilizes electricity–pulsed low-amperage, high-voltage direct current–to shock crayfish into the water column in front of the trawl unit (Cain et al. 1986) (Figure 20). The beam trawls can be used to some degree in vegetation as they can push it down and out of the way as they pass. Once vegetation dies, it breaks and clogs the trawl so that harvesting lanes are required then. The electric trawl is mounted on a boat. Other systems utilize wheel-driven, light-weight propulsion units (Bergeron 1987; Riche 1989). Hydraulic fluid is the preferred way to drive the wheels (Figure 21).

Other automated harvesting devices include a linear move irrigation structure that was designed to periodically move traps across a pond, set them, empty them into a collecting system, and move them again (Morgan et al. 1982-83). It has yet to be used on a commercial scale. Two others are submerged plastic irrigation pipe to move crayfish to a centralized holding point. One unit has traps attached to it with a funnel from each trap into the pipe. A pump moves water through the pipe which carries the crayfish to the

2. All "reported" patents are not referenced in this text as they are not available to the author.

Figure 19. A Crawfish Combine with a stern-mounted, hydraulic propulsion wheel. J. Huner.

collecting unit. It is said to be patented (1990, M. LeBlanc, Manager, University of Southwestern Louisiana Experimental Farm, Lafayette, Louisiana, USA, personal communication). The other unit has openings directly into the pipe itself. The crayfish enter these and are moved to the collection point by pistons attached to a cable. The cable-piston unit is activated at various intervals to move the crayfish in the pipe to the collection point. It is also reported to be patented (1990, J. Smith, Jennings, Louisiana, USA, personal communication).

In general, it appears that automated harvesting could be an important contribution to the crayfish industry but information on management methods such as harvesting intensity, time of day to harvest, use of baits and other attractants to concentrate crayfish, etc. is needed before they can become widely used.

Orconectes spp. may be harvested with traps and seines (Somers and Stechey 1986; Rach and Bills 1987; Brown et al. 1989; McCartney and Garrett 1989; Huner 1990a). Both fish and manufactured baits are effective (Somers and Stechey 1986; Rach and Bills 1987;

Figure 20. The Cain et al. (1986) automated crayfish harvester which uses electrical current to induce the crayfish to flip upward into the path of the bow-mounted aluminum catch unit. J. Huner.

Brown et al. 1989). Density and temperature (Somers and Stechey 1986; Brown et al. 1989) directly effect catch-per-unit effort.

Fixed, unbaited "pound net" type traps are used to capture crayfish in Spain (Huner 1990a, d). These take advantage of movement of crayfish in response to water flow. Traps are actually cylindrical eel traps (fyke, or hoop, nets) with guide panels at their entrances. Similar, experimental wire-mesh units have shown promise for use in harvesting crayfish in Louisiana studies (Romaire 1989b).

Many crayfish are stranded when ponds are drained, especially in April and May for rice cultivation. They are either unable or unwilling to burrow into the pond bottoms. Some farmers secure modified shrimp trawls to drain structures or make temporary box traps for the drains to catch crayfish when ponds are drained (Huner, unpublished observations). Crayfish may follow water out of ponds at night but there is not enough experience with this technique to predict success or failure. It is not unusual for the crayfish to remain in the pond and not follow the water out. The crayfish may move into a stream of fresh water entering such fields and can occasionally be collected in large numbers where the water enters the fields.

Figure 21. The Crawbine automated crayfish harvester with aluminum catch unit raised to dump crayfish into a holding unit under the operator's seat. J. Huner.

There is no commercially acceptable alternative to using traps to harvest crayfish. Automated devices that would significantly increase harvest efficiencies would be a boon to the industry.

Predaceous and Nuisance Organisms

Crayfishes are consumed by all manner of carnivorous invertebrates including other crayfish and vertebrates. Crayfish ponds are managed to exclude predaceous fishes (Huner 1990d). This permits unchecked proliferation of large predaceous insects including odonate nymphs and the hemipterans (Barr et al. 1978; Huner and Naqvi 1984). Laboratory and exclosure studies in the field show that these organisms prey heavily on confined crayfishes as long as the size differential is great but their role in ponds is not clear (Dye and Jones 1975; Witzig et al. 1986). These arthropods normally colonize at the same time that ponds are flooded (Witzig et al. 1986). As a result, they grow at about the same rate as the young crayfish and are unable to prey on those crayfish. One study demonstrated that the large odonate nymph, *Anax junius*, would readily prey on small *P. clarkii* in the laboratory (Witzig et al. 1986) but repeated dissections of *A. junius* nymphs caught in ponds with high crayfish populations revealed no evidence that they had consumed crayfish (Gaude 1982).

It is possible to project the natural mortality that can be attributed to predaceous arthropods in ponds based on review data on mortality presented by Romaire and Lutz (1990). Mortality in pool studies where few predaceous arthropods occur has been 1-4% per week for young-of-the-year *P. clarkii* and *P. zonangulus*. Mortality in earthen ponds with good water quality and no predaceous fishes has been 2-13%. Therefore, mortality from predaceous arthropods (including crayfish!) likely ranges from 1-9%.

Even if clear evidence is generated showing that various invertebrate predators negatively impact crayfish survival and growth, there is apparently little that can be done about it. Pesticides used to kill the predators will kill the crayfish, too. Treatment of pond surfaces with oil materials to selectively kill air breathing hemipterans presents several problems. The toxicant may kill crayfish if they come to the surface to use atmospheric oxygen and/or the toxicant may impart a disagreeable odor/taste in the crayfish. Furthermore, emergent vegetation might render the treatment ineffective.

The corixid bugs (Hemiptera) called "water boatmen" create a crayfish quality control problem especially in the spring in devegetated ponds (Huner 1990c). They lay eggs on newly molted crayfish. Numbers can be so high that the crayfish appear unsightly although there is little evidence that they damage the crayfish. This can be a serious problem when crayfish are rejected by buyers or a low price is paid for them. Control methods such as oiling ponds (see above) may work but no research has been done to determine the benefits of the treatment and impact on crayfish. Mosquitofish, *Gambusia affinis*, and fingerling sunfishes, (centrachidae), when present in high numbers, appear to control water boatmen populations very effectively but management recommendations are not available.

All types of aquatic reptiles, amphibians, and fishes prey upon and consume crayfish and Penn (1950) provides a lengthy list of such predators. Crocker and Barr (1968) summarized much of the existing information on homeothermic and poikilothermic verte-brate crayfish predators.

Fishes are the most significant predators encountered in crayfish ponds (Huner and Barr 1991). Hardy species like the black bull-head, *Ictalurus melas*, and the green sunfish, *Lepomis cyanellus*, thrive in crayfish ponds, being well adapted to life in unstable aquatic ecosystems. They can be especially bothersome if ponds are not properly drained during the summer as they have very high reproductive rates and survival rates in conditions that kill most other fishes. It is imperative that all standing water be eliminated from crayfish ponds when they are drained or reduced as much as possible followed by application, at recommended rates, of the fish toxicant rotenone. Rotenone should be applied prior to pond flood-ing if standing water is present.

Bullfrogs, *Rana catesbeiana*, and amphiuma, *Amphiuma means*, are well-known amphibian predators of crayfish (Crocker and Barr 1968). Both are common around crayfish ponds.

Turtles, especially the red ear turtle, *Pseudmys scripta*, can de-stroy many trapped crayfish by consuming abdomens that protrude through the trap mesh (Huner 1986b). These reptiles can be an especially noisome problem in the spring in low-lying areas sur-rounded by surface waters. Alligators, *Alligator mississippiensis*, especially small ones, consume large quantities of crayfish (Crocker and Barr 1968; McNease and Joanen 1977).

The aquatic rodents, nutria, *Myocaster coypu*, and the muskrat, *Ondatra ziebethicus*, are normally herbivorous but create real prob-lems by burrowing through levees and causing collapses (Huner and Barr 1991). Predaceous, semi-aquatic mammals like mink, *Mustella vison*, raccoon, *Procyon lotor*, opossum, *Didelphis marsu-pialis*, and otter, *Lutra canadensis*, do prey on crayfish but are generally not abundant enough to consume many crayfish. Howev-er, otters may damage traps to get both bait and crayfish during winter months. *Procambarus clarkii* has become an important food resource for the European otter, *Lutra lutra*, after its introduction in Spain in the early 1970s (Delibes and Adrian 1987).

All manner of birds consume crayfishes (Huner and Abraham 1983; Martin and Hamilton 1985; Huner and Barr 1991). Various species not thought of as predators include blackbirds, grackles, and crows, as well as barred owls and some hawks. Piscivorous ducks, especially mergansers, and cormorants readily eat crayfish. All herons, egrets, and ibises consume crayfish but those of greatest concern are yellow-crowned night herons, *Nyctanassa violacea*, white ibises, *Eudocimus albus*, and great (common) egrets, *Casmeroidius albus*. Wading birds are especially common when crayfish are abundant and/or vulnerable such as times when oxygen levels are low, ponds are being flooded, or ponds are being drained.

It has not been possible to quantitate the impact of avian predators on crayfish production (Huner and Abraham 1983; Martin and Hamilton 1985). Ponds have the potential to produce up to 3000 kg/ha but the average is 500-600 kg/ha. It is probable that when ponds are drained and/or flooded avian predation can cause enough mortality of broodstock to reduce the numbers of young needed to generate maximum production. However, other factors such as low harvesting intensity as influenced by market conditions, poor water quality especially when ponds are flooded, and forage deficiencies can account for less than optimal conditions for production (Huner 1990b).

Polyculture

Unintentional polyculture of fish and crayfish has a long history in North America (Huner 1976). Tertiary burrowers like *O. immunis*, *P. clarkii*, and the white river crayfishes are commonly found in ponds used to cultivate fingerlings of species that, as adults, prey heavily on crayfishes, or to cultivate non-carnivorous fishes like cyprinid bait minnows. Fish farmers usually try to avoid crayfish because they prey on eggs on spawning mats, compete for space and food, and interfere with harvesting by killing small fish in nets. Management actions to counter these problems include not filling ponds until just before fish are stocked and/or using toxicants just before fish are stocked (Huner 1976; Bills and Marking 1988). Leaving ponds flooded during winter months but stocking with predaceous brood fish has been shown to be an effective way to keep crayfish populations under control (Rach and Bills 1989).

Food fishes have been stocked experimentally into deeper crayfish ponds and grown during the normal spring, summer, autumn growing period (Green et al. 1979). This was done after most of the crayfish had become too large to be effectively preyed upon by the species used or after most of the crayfish had been harvested. Species that have been cultivated with *P. clarkii* in ponds include channel catfish, *Ictalurus punctatus* (fingerling and food fish), hybrid buffalofish, *Ictiobus* spp., golden shiners, *Notemigonius crysoleucas*, and paddlefish, *Polydon spathula*, as well as freshwater prawns, *Macrobrachium rosenbergii* (Green et al. 1979; Huner et al. 1983a, b; Cange et al. 1986b; Huner 1986a). Brood crayfish constructed burrows in pond levees and generally did not emerge until after the autumn harvest of fish which involved complete pond draining. However, crayfish hatched in late spring were harvested in mid-summer with yields up to 500 kg/ha in freshwater prawn-channel catfish studies (Huner et al. 1983a, b).

A problem with the preceding management scenario was that no forage vegetation could be grown during the summer because ponds were full. Later studies demonstrated the feasibility of draining ponds in May and planting rice as a forage for crayfish in the upper, shallower, one half to one third of the pond. Small fish were stocked in the deeper, refilled end and water level was raised with the rice growing upward in response to rising water levels. Fish were harvested in early autumn by draining the ponds and ponds were refilled for crayfish. The freshwater prawn, *M. rosenbergii*, and tilapia fish, *Sarotherodon* spp., were incorporated into this management scheme with channel catfish (1990, J. W. Avault, Jr., Louisiana State University, Baton Rouge, Louisiana, USA, personal communication).

Fingerling fishes such as channel catfish, goldfish (*Carassius auratus*), koi carp, *(Cyprinus carpio)* etc., can be grown in crawfish ponds during the March-May period as the ponds have rich zooplankton fauna, ideal food for such fish (Huner and LeBlanc 1991). Weir-type traps, such as those used in Finland to catch fingerling fishes, will need to be developed to recover such fish when ponds are drained.

None of the experimental polyculture scenarios has been practiced commercially in the USA. The financial situation has not favored such endeavors but they are at least biologically feasible.

Crayfish, *P. clarkii,* are routinely harvested from Chinese carp poly-culture ponds in the People's Republic of China (Xinya 1988).

A unique form of polyculture involves management of ponds for waterfowl hunting (Perry et al. 1970; Huner and Barr 1991; Nassar et al. 1991). Waterfowl are readily attracted to crayfish ponds because they are shallow and there is quite a lot of feed in the form of weed seed and unharvested rice grains. Leasing of hunting rights can be very lucrative in areas where waterfowl are abundant.

In some places, swampy areas are leveed and diked so that water can be trapped in the autumn and released in the spring. Trees provide seed, especially oak, *Quercus* spp., that attracts ducks. These are called Green Tree Reservoirs. Some crayfish production may be realized in these, though they are usually drained earlier in the spring than is conducive to good crayfish production. This reduces stress on the trees so that they will thrive and continue to generate seed to attract ducks in the next year.

Transportation

Crayfish are normally transported in open mesh, plastic vegetable sacks which hold 16-23 kg of crayfish, depending on size (Huner and Barr 1991) (Figure 22). These sacks are convenient and inexpensive but they are not especially effective in protecting the crayfish from physical damage associated with moving sacks and storing them vertically. Sacks are not effective in protecting the crayfish from heat and cold or dehydration during transportation. As a result, farmers and processors are gradually moving to use of nesting or collapsing plastic boxes for transporting crayfish (Huner 1990d) (Figure 23). While some boxes have to be adapted for use with crayfish (by making drain holes in the bottoms because crayfish can suffocate in the water), the crayfish are protected from physical damage other than that which they might do to themselves. However, graded crayfish are less likely to damage themselves because the size differential is small in any particular grade. Hard containers are 10-20 times more expensive than sacks which are discarded. Therefore, they must be returned after use or replaced. Open mesh bags must be covered in transit and during prolonged outdoor storage to protect crayfish from extreme temperature changes and dehydration.

Figure 22. Stacks of vegetable sacks filled with crayfish awaiting processing or trans-shipment. J. Huner.

There has been no published research on transportation methods in the crayfish industry. The methods have developed within the industry. They have served well for transportation of crayfish to local markets within Louisiana or to processing plants. However, the development of out-of-state markets and the need to move larger volumes of crayfish from pond to pond for stocking purposes suggest a need to re-examine transportation methods. For example,

Figure 23. On-boat grader and plastic boxes. Boxes may ultimately replace traditional vegetable sacks currently used to hold and transport crayfish. J. Huner.

there is much to be gained in terms of enhanced survival of both broodstock and food crayfish by describing the physiological changes occurring during transport and associated aerial exposure.

Market Development

Louisiana produces over 90% of the U.S. supply of crayfish and consumes 70-90% of it (Huner 1989c; Roberts and Dellenbarger 1989). The remainder of the crayfish is produced in the Southeast, the upper Midwest, and the Far West (Huner 1989c). The crayfish produced in the Southeast is consumed locally. Most of the crayfish produced in the Far West is exported to Europe, primarily Scandinavia. Small crayfish produced in the Midwest is sold for fish bait and larger crayfish are exported to Europe.

It is critical to the Louisiana industry to develop stronger markets outside of the state. Development of such markets is adversely impacted by seasonality of supply, poor image, and high cost of product, especially meat, relative to competitive crustacean products (Dellenbarger et al. 1986, 1990).

There are two sources of crayfish in Louisiana, wild caught and farm-raised (Roberts and Dellenbarger 1989). The wild caught crayfish come primarily from the Atchafalaya Basin Floodway in the south-central region of the state. Most are harvested during the months of April, May, and June (Dellenbarger et al. 1986). However, the magnitude of the crop is dependent upon flood waters received from the Mississippi River (Dellenbarger and Luzar 1988; Huner 1990d). (The Atchafalaya Basin Floodway is the principal distributary for the Mississippi River.)

Water levels cannot be predicted for more than a few weeks in advance. When waters are low, fishermen simply have no access to the crayfish. When waters are high, the volume can be so high that prices plummet (Dellenbarger and Luzar 1988). Processors must be very cautious about making major purchases of live product for peeling lest they be forced to hold high cost inventories of processed meat. Dellenbarger and Luzar (1988) developed a mathematical model to predict the effects of water levels in the Atchafalaya Basin Floodway on crayfish supplies and prices. The value of this model is marginal because of the unstable nature of the water levels in the Mississippi River system. Long-term projections simply cannot be made.

It is clear that markets for crayfish do exist throughout the USA especially in areas with a tradition of consuming seafood (Roberts and Dellenbarger 1989). However, desired product forms vary significantly across the country according to market surveys (Dellenbarger et al. 1986, 1990, 1990-91). Aggressive marketing programs are necessary to garner greater market share. The development of frozen products including whole cooked crayfish, crayfish meat, prepared entrees, and soft-shell crayfish now permits establishment of the inventories necessary to provide the stable supplies necessary to make marketing programs successful.

Economics

Excellent reviews of crayfish culture economics are available for Louisiana (Dellenbarger et al. 1987) and South Carolina (Pomeroy et al. 1989) crayfish culture operations. These reviews demonstrate several dramatic facts. First, economies of scale show that larger units are more successful than smaller ones. Second, integration of crayfish culture into a rotation with agronomic crops can be more profitable than crayfish monoculture because of shared, common expenses for pumps, water control structures, etc. Third, based on average prices paid for crayfish since 1987 and adjusting them for inflation, most crayfish ventures are operating at a loss with industry-wide average production levels of 500-600 kg/ha. Finally, harvest costs, labor, and bait account for 60-80% of annual operating costs. Data on the break-even prices needed for various crayfish operations based on the bulletin of Dellenbarger et al. (1987) are provided in Table 11. The economies of scale are very clearly presented. Note that the lowest production level shown is greater than the state's average production! Information on fixed and variable costs for a 16.7 ha crayfish pond is presented in Tables 12 and 13. Note that land ownership is assumed. While dated, these data provide an excellent outline for anyone contemplating *Procambarus* spp. culture in earthen ponds.

Economic conditions were poor in the Louisiana and the Texas crayfish industries (Huner 1989d) for the last two seasons of the 1980s. This was compounded by the sudden importation of large quantities of low price *P. clarkii* meat from the People's Republic of China in late 1991. Pond area declined about 50% in Texas and at least 10%, if not more, in Louisiana. This followed approximately 25 years of continuous expansion in Louisiana and at least 10 years of expansion in Texas. The apparent cause of the decline has been declining prices. The region experienced a hard economic blow from decline in oil revenues during the period as those who traditionally bought crayfish and crayfish meat were no longer able to do so, at least at past levels. In the absence of established, dependable, out-of-state markets, prices fell.

Crayfish producers who continue to practice good husbandry and who reduce production costs are able to operate profitably. Higher

Table 11. Break-Even Prices Required for Various Southwest Louisiana Crayfish Farm Operations, After Dellenbarger et al. (1987).

A. Crawfish Monoculture

Acreage	Production in kg/ha				
	795	1,023	1,250	1,977	1,705
	— dollars($) /kg —				
10	4.81	3.26	2.66	1.24	1.96
20	2.75	2.13	1.74	1.50	1.30
40	2.07	1.61	1.32	1.12	0.97
80	1.74	1.34	1.10	0.92	0.81

B. Crawfish/Rice Polyculture

Acreage	Production in kg/ha				
	795	1,023	1,250	1,977	1,705
	— dollars($) /kg —				
10	3.63	2.83	2.31	1.96	1.69
20	2.46	1.89	1.56	1.32	1.14
40	1.89	1.47	1.21	1.03	0.88
80	1.67	1.10	1.08	0.90	0.77

prices for larger crayfish mean that some producers with low production can still generate a net profit. However, all segments of the industry must work together to insure that the crayfish industry operates profitably. It could easily expand by factors of 3 or 4 if additional market demand could be generated.

Semi-Intensive/Intensive Culture

Crayfish may be raised through many generations in closed systems (Black and Huner 1975). This is not done commercially in North America because of the expense involved compared to the value of the product. Any recirculation system may be used. Those described by Malone and Burden (1988) in their inexpensive, highly detailed design manual for soft-shell crayfish production would be more than adequate for the reader regardless of skill level.

Table 12. The Estimated Investment Costs and Annual Depreciation Charges for a 16.2 ha Monoculture Crayfish Pond in Southwestern Louisiana in 1987, After Dellenbarger et al. (1987) and de la Bretonne and Romaire (1989b).

Item	Investment/ha Dollars ($)	Depreciation/ha Dollars ($)
Pond Construction		
Earth Moving	387	
Water Control Structures	91	
Ground Cover	21	
Total	499	
Stock and Equipment		
Broodstock	51	5.12
Well	688	34.38
Pump	608	40.56
Engine-Gearhead	786	52.38
Oxygen Meter	38	9.38
Crawfish Combine	312	31.25
Truck	563	93.75
Traps	387	129.00
Cooler	75	14.94
Scale	6	1.19
Aerator	27	2.63
Mower	44	14.56
Waderes	15	7.00
Total	3,600	436.14
Grand Totals	4,099	436.14

Table 13. Estimated Annual Operating Costs Associated with a 16.2 ha Monoculture Crayfish Pond in Southwestern Louisiana in 1987, After Dellenbarger et al.
(1987) and de la Bretonne and Romaire (1989b).

Variable Costs	Dollars ($) / ha
Forage	102
Fuel	115
Repairs and Maintenance	66
Labor @ $5/hour	182
Herbicides	10
Sacks	10
Bait @ $.35/kg	365
Total	850
Fixed Costs	**Dollars ($) / ha**
Depreciation	436
Interest (12%)	255
Total	661
Grand Total	1,511

Soft-shell crayfish production units are intensive culture units
(Huner 1988c; Malone and Burden 1988; Culley and Duobinis-
Gray 1990); however, production of soft-shell crayfish is semi-intensive because the immature crayfish used are produced in crayfish
ponds or are caught in natural fisheries. (Mature crayfish do not
molt regularly and are not used in production of soft-shell crayfish
for food.) These crayfish are held at densities of about 220-250 per
square meter and are fed until they near molt, a time when they
cease to feed (Figure 24). They are then removed from the feeding
unit and transferred to a molting unit where they will not be at-
tacked by non-molting crayfish (Figure 25). Premolt crayfish removed from feeding trays are constantly replaced by new crayfish
to maintain production level densities.

Figure 24. A Louisiana soft-shell crayfish production system with banks of production trays. J. Huner.

Production units, called trays, are roughly 1 m × 2.5 m × 0.15 m deep (Huner 1988c; Malone and Burden 1988; Culley and Duobinis-Gray 1990). Water levels are held at about 2.5-3.0 cm in feeding trays, so that crayfish can utilize atmospheric oxygen if there is a system malfunction. The water level in molting trays must be deeper or the crayfish may be misshapen. Target water quality parameters are: pH, 7.0-8.0; DO, over 5 ppm; nitrite, less than 0.3 ppm; ammonia, less than 0.3 ppm; and temperature, 24-26°C.

System turnover time, the period over which all crayfish stocked initially die or molt, is 30-40 days (Culley et al. 1985; Culley and Duobinis-Gray 1987a, b). Good production is considered to be a 3-4% molting rate per day. Ironically, crayfish molt regularly in ponds on 7-20 day cycles even at lower ambient temperatures (Hun-

Figure 25. A close-up of crayfish in a soft-shell crayfish production tray. J. Huner.

er 1978b). At much lower densities, about 40 per square meter, somewhat smaller, individually held *P. clarkii* routinely molt within 10 days of being placed into holding systems after capture in ponds (Huner and Avault 1976a). Furthermore, when crayfish eyestalks are ablated, most intermolt or early premolt crayfish molt in 8-10 days after being placed into soft-shell crayfish production units (Huner and Avault 1977; Huner et al. 1990; Chen et al. 1992). Thus, considerable effort is being devoted to eliminating the inhibitory factor(s) that delays molting so much in conventional soft-shell crayfish systems.

The soft-shell crayfish production technology utilized in Louisiana is referred to as the "Culley System" after its developer, Dudley D. Culley. It is described by Culley and Duobinis-Gray (1990) and explained above. Another system, the Bodker System, developed by

Edward Bodker (Bodker 1984), has not found widespread acceptance in the industry but offers a unique alternative that may eventually lead to its more widespread use. This patented method utilizes
much deeper units, with water depths in excess of 15 cm. The units
have a gradual slope leading from the bottom of one end to the water
surface at the other end with a shallow lip there. "Habitats"
constructed from units of PVC pipes are placed in the deep end.
After crayfish are loaded, non-molting crayfish remain in the deeper
water habitats while late premolt crayfish move into the shallow
water where they may be easily collected and moved into a molting
tank.

Another patented soft-shell crayfish production system is in commercial development. The so-called soft-shell crayfish "separator"
(Malone and Culley 1988) is a conventional tray with a continuous
inner maze through which a constant water flow is maintained.
Non-molting crayfish orient toward the current but molting crayfish
are unable to maintain their positions and are carried through the
maze to a collection area. They exit through an electrified gate
carried through it by flowing water. Non-molting crayfish will normally not pass through the gate. The obvious advantage of this
system is that it reduces labor substantially. Units may be stacked
seven or more high which reduces the cost of facilities.

Details about operation of the "separator" are now being published. Robin et al. (1991) present the overall operational and management strategies for such units emphasizing the importance of
having a controlled source of intermolt crayfish for loading commercial systems. This assures a constant supply of crayfish for
stocking units with 10 vertically stacked production trays. Rondelle
et al. (1991) describe a top loading system for supplying crayfish to
the top tray and returning escaping crayfish to the system. Density
of about 230 per square meter are achieved through natural movements of the crayfish from one tray to the next. Electrical gates
effectively stop further movements once the design density has
been reached. Malone et al. (1991) provide a description of a prop-
washed polyurethane bead biofiltration system that supports the
commercial stacked tray system described by Robin et al. (1991).
This is far more efficient than the static and up-flow sand filters

previously recommended for support of recirculating soft-shell crayfish systems (Malone and Burden 1988).

While recirculating systems are recommended for soft-shell crayfish production, flow-through systems have been successfully used (Huner 1988c). Cost of heating water is greater than that of recirculating systems and great volumes of water must be discarded generating concerns about waste disposal permits. A definite advantage of flow-through systems is that there are no water quality problems when a good water source is available.

Soft-shell crayfish are produced primarily for fish bait outside of the lower South (Calala 1976; Huner 1988c, 1990a). The problem of molt inhibition is resolved by using only late premolt crayfish that will molt in 1-3 days. They are often captured using seines in small, obstruction-free ponds adjacent to the production units (Figures 26 and 27). This is generally not practical in the lower South where crayfish are harvested with traps and vegetation generally makes it impossible to seine ponds effectively. However, Culley et al. (1985) noted that daily molt rate almost doubled when unsorted, seine-caught crayfish were used experimentally in a Louisiana soft-shell crayfish system. Should automated trawl harvesting systems become effective, a source of late premolt crayfish would become available. These, however, are very sensitive to handling stress and molting facilities would have to be nearby the source.

An advantage that soft-shell crayfish producers in the Great Lakes region have over competitors in other areas is that mature male and female crayfish molt more or less synchronously twice each growing season (Payne 1978). Time of molting is controlled by temperature and reproductive status. Males molt before females in the spring or early summer because the females will not molt until they have released their young. Males molt again in late summer/early autumn and females molt after them. This situation makes it possible to capture them in the early to mid-premolt period after they are committed to the molt. All will molt within about 10 days of capture if the timing of trapping is correct. The producer can follow the molt from south to north during the molting periods to obtain animals from natural populations.

The Louisiana tray system of soft-shell crayfish production is well described in published bulletins by Culley and Duobinis-

Figure 26. An open-air soft-shell crayfish system in Ohio. Water flows from a pond on one side of a levee, through the unit and into a pond on the far side. Premolt crayfish, *Orconectes* spp., are caught with seines in the ponds and transferred into the molting units. J. Huner.

Gray (1990) and Malone and Burden (1988). Information about economics of that system is presented in Tables 14 and 15. The northern system of soft-shell crayfish production is well described by the manual of Calala (1976).

Crayfish hatchery systems have no real importance in North America (Huner and Gaudé 1989) but two are described here to show their simplicity. In the Black system (Black and Huner 1975), *P. clarkii* are grown individually in 25 cm diameter glass stacking bowls 7.5-10.0 cm deep. A layer of aquarium gravel is maintained over the bottom and the crayfish are fed periodically with bits of dried pet food. Green vascular aquatic vegetation (elodea being preferred) and well-composted hardwood leaves are maintained continuously as additional food sources. Water is changed once a week. Once the crayfish mature at sizes of 6-7 cm total length after

Figure 27. A close-up of molting trays of an Ohio soft-shell crayfish production system. J. Huner.

3-4 months at room temperature, around 22°C, a mature male is placed in a bowl containing a mature female. After copulation is confirmed, the male is removed. The female will usually lay eggs in 6-8 weeks with incubation and release of young being completed in another 3-4 weeks.

   In the Gooch method, mature female *P. clarkii* are captured in ponds in late April or May and placed into hatchery units described by Trimble and Gaudé (1988). These units consist of pieces of 10 cm × 10 cm plastic sewer pipe placed in shallow wooden trays or, they may even be placed in plastic cafeteria trays as long as they will hold up to 3 cm of water (Figure 28). A screen top is placed over each unit to prevent escape. Water is added periodically but the crayfish are not fed. Most mortality occurs within the first two weeks. Dead crayfish must be removed to prevent common water within the unit from fouling. The units are kept in a relatively cool place, 20-22°C, with subdued lighting. Females are assumed to

Table 14. Estimated Investment Requirements and Depreciation Charges for Soft-Shell Crayfish Production Systems [500 kg capacity] in Louisiana in 1988, After Caffey (1988).

Item	Flow-Through System		Recirculating System	
	— Dollars ($) —			
	Investment	Depreciation	Investment	Depreciation
Greenhouse	4,990	490	4,990	490
Limestone Slab	345	17.50	345	17.50
Water Well	2,665	266.50	1,700	170
Plumbing	318	31.80	——	——
BioFiltration Unit	——	——	3,485	348.50
Sump and Reservoir	——	——	350	17.50
Wiring	100	5	100	5
Gas Line	150	7.50	150	7.50
Trays (44)	1,660	415	1,660	415
Stands	572	190.66	572	190.66
Refrigerator/ Freezer	600	85.71	600	85.71
Double Sink	55	3.60	55	3.60
Desks, Tables Chairs	50	10	50	10
Water Heating (250,000 BTU boiler)	1,330	56.60	——	——
Water Heating (water heater)	——	——	300	30
Miscellaneous	619	——	709	——
Totals	13,254	1,600	15,066	1,792

Table 15. Estimated Annual Operating Costs for Flow-Through and Recirculating Soft-Shell Crayfish [500 kg capacity] Production Systems in Louisiana in 1988, After Caffey (1988).

Item	Flow-Through	Recirculating
	— Dollars ($) —	
Operating Expenses		
Labor	4,200	4,200
Owner/Operator (part-time)	1,008	1,008
Crawfish		
Initial Stocking	1,000	1,008
Monthly Replacement	6,048	6,048
Transportation	510	510
Electricity	1,065	500
Water Heating	5,075	300
Feed	330	330
Miscellaneous	700	700
Total	19,936	14,596
Fixed Costs		
Interest	1,686	1,765
Depreciation	1,600	1,792
Total	3,286	3,557
Grand Total — Expenses	23,222	18,153

RE: Caffey, Rex H. 1988. An Economic Analysis of Alternative Softshell Crawfish Production Facilities. Paper prepared to presentation at the Undergraduate Papers Session, 1988 American Agricultural Economics Association Annual Meeting, Knoxville, Tennessee, July 31 - August 3, 1988.

Figure 28. A stacking crayfish hatchery system for *Procambarus clarkii.* Mature and, presumably, mated female crayfish are placed individually in each compartment made of pieces of plastic pipe. J. Huner.

have been mated when caught. A representative number will lay eggs during the September-October period and young will be available for use about 4 weeks after eggs are laid. This is certainly a very simple method that may have some usefulness in future years. A modification of the method involves the placement of newly mature females into such a system during other times of the year. This approach has shown that it is possible to generate young at other times (Huner, unpublished data).

No literature exists about hatchery production of white river crayfishes. Methods that worked well with *P. clarkii* generated very erratic results with *P. zonangulus* (Huner, unpublished data). Bovbjerg (1956) describes a colonial system for cultivating *Procambarus alleni* in large tanks. This has worked for *P. clarkii* (personal observation).

Species like *O. immunis*, *O. rusticus*, and *O. virilis* will spontaneously spawn in the spring as water temperatures warm (Aiken 1969; Berrill and Arsenault 1984). Young may be produced in aquaria if mature males and females are obtained during the late summer-early autumn and held over winter at ambient temperatures and photoperiods. Spawning may be induced earlier in the spring if temperatures are raised (Aiken 1969). Leonhard (1981) describes a system that is generally suitable for use with any of the *Orconectes* spp. of commercial importance. Brown et al. (1990) recently reported production of many young *O. virilis* by holding mature crayfish outdoors in large metal pools over the winter in central Illinois.

Growing small crayfish to useable size for fish bait or food in intensive systems is not cost effective. Pool studies indicate that densities of fewer than 10 and 30 per square meter are necessary to generate food and bait-sized crayfish in a reasonable period of time. Final biomass at such densities would be less than 0.25 kg per square meter (Smitherman et al. 1967; Clark et al. 1975; Goyert and Avault 1978, 1979; Lutz and Wolters 1986).

There is clear interest in crayfish for sale as pets. There is no need for confined culture of normal crayfish for this market but blue and white mutations are found and are apparently popular (Black and Huner 1980; Tabrosky 1982; Quinn 1989). These do not perpetuate well in pond culture so anyone contemplating production would want to use intensive culture methods. Larger crayfishes will disrupt aquaria and consume expensive ornamental fishes. However, they are interesting and welcome pets if proper attention is given to other aquarium residents. A major problem with use of crayfish as pets involves the disposal of an unwanted, living specimen. Many of the unwanted introductions of crayfishes around the world have come as a result of a kind- hearted pet owner who did not kill unwanted crayfish before disposing of them (Hobbs et al. 1989).

Genetical Considerations

Crayfish genetics have been little studied. Selective breeding has been limited to examination of simple recessive mutations involving body color and eye pigmentation (Black and Huner 1980) and determination of heritabilities of certain desirable commercial traits (Craig and Wolters 1988; Lutz and Wolters 1989). Heritability values were low compared to those reported for other animal taxa. Electrophoretic studies of enzymes also show low levels of heterozygosity (Busack 1988, 1989). A study by Busack (1988) examined heterozygosity of *P. clarkii* and of what was then identified as *P. a. acutus* in populations from areas in the USA. The *P. a. acutus* populations exhibited low heterozygosity when compared to other animal taxa but much greater heterozygosity than the *P. clarkii* populations. It should be no surprise, then, that *P. a. acutus* has been subdivided into several separate species (Hobbs and Hobbs 1990).

Mitochondrial DNA offers a useful tool for distinguishing between species and identifying crayfish stocks (Palva and Huner 1989). However, crayfish tissues have not lent themselves to the rapid mitochondrial DNA isolation methods used with fishes so time and costs involved in such studies are very high. As a result, crayfish stock studies are currently limited to protein electrophoresis although rapid advances are being made in mitochondrial DNA analysis methods.

Studies of crayfish karyology are complicated by the large number of chromosome pairs. *Procambarus clarkii* for example has a 2N number of 188 (Murofushi et al. 1984).

Crayfish genetics is clearly an open field. Low levels of heterozygosity suggest that more immediate commercial gain is to be realized by controlling environmental conditions than selecting for a commercially superior crayfish. The value of a genetically superior crayfish is questionable as long as crayfish are cultivated by establishing sustaining populations in earthen ponds from which "wild" crayfish cannot be excluded.

Orconectes *Culture*

Interest in culture of crayfish for food in the central and northeastern USA has increased as interest in aquaculture in general has increased in these regions (Brown et al. 1989, 1990). Various *Orconectes* spp. that grow to suitable size for food include *O. immunis, O. limosus, O. nais, O. rusticus,* and *O. virilis.* With the exception of *O. immunis,* a tertiary burrower, these crayfishes have wider bodies and chelae than burrowing species (Figure 29). Growth is greatly dependent on availability of food and population density (Hazlett and Rittschof 1985; Klaassen 1986; Momot 1988; McCartney and Garrett 1989). It should be noted, however, that both *P. clarkii* and white river crayfishes, *Procambarus* sp. and *P. a. acutus*, can be locally abundant in the region and should not be ignored as cultural candidates where it is permissible to cultivate them. In addition, *O. longedigitus*, is an especially large species found in rivers and streams in Missouri. The species seems to have adapted well to impoundments created by dams in the region and apparently has potential for pond culture (Davila and Wilkinson 1990).

Interest in *Orconectes* spp. culture exists in eastern Canada. Crocker and Barr (1968) reported that *O. immunis* was grown for bait in Ontario. Stechey and Somers (1983) discuss the merits of several species for culture in Ontario.

Only *O. immunis* is a species that inhabits temporary habitats and can, as a result, be cultivated in semi-permanent ponds. Unless perpetuating populations of other species are established in permanent ponds, a source of young must be developed for restocking growout ponds each spring. Klaassen (1986) and Brown et al. (1990) have explored these alternatives with *O. nais* and *O. virilis*. Free reproduction in ponds leads to higher population densities and reduced growth rates, often with maturation at smaller sizes. Crayfish acceptable for food markets can be produced in one summer if young of the year are stocked at relatively low densities. However, acceptable, food-grade crayfish can be produced in 12-14 months if young crayfish are carried over the winter in the same pond at somewhat higher densities. Actual control of numbers can be achieved only if the number placed into the pond during the first summer is controlled. Removal of all crayfish and restocking is necessary in the second summer if overpopulation and stunting are to be avoided. Rotations using several ponds including one as a nursery pond are recommended. Bait-sized *Orconectes* spp. can be produced in one summer.

Langlois (1935) reported production of *O. rusticus* exceeding 850 kg/ha in bass (*Micropterus* sp.), fingerling ponds in Ohio USA. The broodfish density was not adequate to control crayfish reproduction and fry/fingerling bass were simply not large enough to consume

Figure 29. Form I male rusty crayfish, *Orconectes rusticus,* an abundant species in the north-central USA with excellent aquaculture potential. Robert Pagel.

the crayfish that would ultimately become a primary food item for them! Rickett (1974) made similar observations for *O. nais* in experimental ponds with and without bass.

Huner (1984) reported that *P. clarkii* and white river crayfish, *Procambarus* sp., generated production levels to 1000 kg/ha in an Ohio fish hatchery which had previously had high densities of *O. rusticus.* Apparently, the *Procambarus* spp. populations displaced the *O. rusticus* populations as the species was still common in the drainage stream for the hatchery.

It appears that ponds used for cultivating *Orconectes* spp. and *Procambarus* spp. in higher latitudes should have water in them during the winter to prevent the burrowing species from freezing in their burrows and the other species which burrow less effectively from being frozen. It has apparently been common practice to harvest fish during the summer. The ponds then refill from seepage and rainfall during the following months so that there is water in them during the winter. It is not entirely clear, however, if "non-burrowing" *Orconectes* spp. survive in the pond proper or if brood crayfish migrate into the refilled ponds from adjacent ponds and/or drainage streams.

Forage for crayfish in central USA fish ponds can be established by partial or total drainage of the ponds in the early spring after the danger of freezes has passed (Calala 1976; Huner 1976). Rye grass or wheat is planted in ponds at most fish hatcheries to serve as green fertilizer for the fry and fingerling fishes that will be grown in the ponds. Organic fertilizers such as cottonseed meal and animal manures are also added to stimulate intense zooplankton blooms. Thus, if the crayfish are not intentionally poisoned to keep numbers down, conditions are ideal for crayfish production which can be in excess of 500 kg/ha. (Note: established burrowing crayfish populations are virtually impossible to eliminate.)

A potentially interesting form of integrated crayfish-agronomic crop culture in the upper midwestern USA involves simultaneous production of wild rice, *Zizaneopsis aquatica* and *O. immunis*. Gunderson (1990, J. Gunderson, Sea Grant College Program, University of Minnesota, Duluth, Minnesota, USA, personal communication) reports that *O. immunis* can be so common as to be a pest in Minnesota wild rice paddies. As a result, research has begun to determine if the two crops can be cultivated with crayfish being "controlled" by commercial harvest.

Forney (1958/revised 1968) published a bulletin dealing with *O. immunis* culture for fish bait in New York State in the northeastern USA. A brief summary of his recommendations follows. The aim of this method is to produce many, small, high-value crayfish actively sought by anglers, especially "value added" soft-shell crayfish. Small, manageable ponds 0.10-0.25 ha are used. Adult crayfish may be stocked in the autumn at 1480-2470/ha. Depth should be 1-2 m and the pond should be filled through the winter and spring. Young crayfish reach fish-bait size by July and are harvested with seines. Unharvested young mature by autumn and form the basic broodstock for the following year. Commercial fertilizer or animal manures may be applied at intervals of 3 weeks to enhance production. The recommended fertilizer/animal manure application rates were 220 kg/ha (6:12:6 or 0:12:0 - NPK) or 49-74 bushels/ha, respectively.

The authors of this text would be remiss if they did not admit some ignorance of crayfish cultural methods that have evolved at fish farms in the central and northeastern USA. There is certainly no real body of literature, other than the references provided above, from which to draw. We therefore encourage those interested in culturing crayfish in those regions to visit commercial and governmental fish farms to assimilate information on methods and suitable species.

A caution about *Orconectes* spp. culture is that there is much concern about widespread introductions of *O. rusticus* in the central USA outside of its native range (Lodge et al. 1985; Momot 1988). While this species has been associated with declines in native crayfish species, changes in sportsfish population structures, and elimination of aquatic macrophytes, opinions about cause and effect vary considerably. Laws and regulations dealing with possession of crayfishes have been passed in some states in the central USA. Therefore, anyone contemplating culture of a particular species should consult local authorities before doing so. Note that *O. limosus* is not especially abundant in the USA but is mentioned as a potential cultural candidate because of its widespread distribution in Europe where it is the most common freshwater crayfish species (Momot 1988). However, it is little used as it has never become popular as a food item.

Culture of Australian Crayfishes, Cherax *spp., in the USA*

There is little experience to date in experimental and commercial cultivation of the Australian crayfishes in the USA. There has been, however, much interest in the subject. Readers are referred to the Australian section of this text for full details of methods but a brief synopsis of the situation in the USA follows.

Entrepreneurs first brought marron, *Cherax tenuimanus*, to the USA as it was the first species widely developed and promoted as an aquaculture candidate in Australia. Its large size encouraged those seeking to produce a freshwater equivalent to the American lobster, *Homarus americanus*. Marron culture has not yet developed much in the USA. Studies by Kartamulia and Rouse (1991) confirm Australian observations that marron does very poorly as water temperatures reach 28-30°C. Marron appear to be best suited for a Mediterranean-type climate which is largely absent from most areas of the USA with sufficient water resources to permit pond culture of the species.

More recently, entrepreneurs have widely touted the red claw, *Cherax quadricarinatus*, as an ideal culture species that would fit in a sales niche between domestic crayfishes and American lobsters. This is a tropical species and research to date in the USA demonstrates that some system for overwintering stock will be necessary in most regions if the species is to be cultivated successfully (Austin 1991; Rouse and Medley 1991).

Summer growth and summer survival of nursed juvenile red claw crayfish (4 g) in earthen ponds were superior to that of juvenile (3 g) red swamp crayfish grown together (Medley et al. 1991b). Survival and final sizes over 165 days were 80 g and 56% for red claw crayfish and 36 g and 16% for red swamp crayfish. Initial stocking rates were 0.5 per square meter for each species. Results for monocultured controls stocked at 1 per square meter were 70 g and 85% for red claw and 38 g and 19% for red swamp crayfish. When hatchling juvenile (0.1 g) red claw were stocked at 10 per square meter, they grew to an average of 41 g with 25% survival (Rouse and Medley 1991). Size ranges for monocultured hatchling and nursed juvenile red claw in the two referenced studies were 3-135 g and 33-140 g.

An economic feasibility and risk analysis study has been conducted for culture of red claw in the southeastern USA (Medley et al. 1991a). The general assumptions upon which this is based are: stock 1 g hatchery-produced juveniles in April at a density of 5 per square meter, grow out period of 190 days, 2000 kg per ha expected yield at an average size of 70 g, hay a primary feed source with a low-protein commercial ration fed supplementally. The basic culture unit is a 1.0 ha earthen pond. Results showed that production costs of 50-100 g crayfish were $11-13 per kg.

The yabby, *Cherax destructor/albidus*, is likely the only Australian species available to producers that can tolerate winter temperatures in open ponds in the southern USA. Although it grows to somewhat larger sizes than red swamp and white river crayfishes (see Australian section for references), it burrows extensively. It is available in the USA but there is apparently no intensive research into developing methods to cultivate the species in the country.

The authors of this section would be remiss if several important factors were not pointed out to those considering cultivation of any Australian crayfish in the USA. First, they are exotic species and it is likely that permits are required to cultivate them in most places. Second, the costs of cultivating them are certain to be high relative to the wholesale values of competing products including other crayfishes, shrimps, and various lobsters. Therefore, it is imperative to consider an investment very carefully before making it. Third, resistance to the crayfish fungus plague, *Aphanomyces astaci*, is yet to be resolved (see Australian section).

Section II:
Freshwater Crayfish Processing

Processing is essential to freshwater crayfish aquaculture and marketing. Historically, the consumption of most crayfish and crayfish products has been restricted to Louisiana and adjacent states including Texas and Florida. Other less significant production and processing occurs in California, Wisconsin, Oregon, and Washington (Huner 1978a).

SEASON

Because of their life cycle complexities, the availability of crayfish to processing plants is highly seasonal. Crayfish from aquaculture sources are available for processing as early as November, while wild stocks are not generally available in large quantity until March. Peak crayfish production occurs in March, April, and May (Huner and Barr 1991). The wavering peaks and valleys in production have significant impacts on processing plant activity and product price (Moody 1989). As production increases from both ponds and wild sources, processing plant activity increases. Generally all processing activity ends by the middle of June when supplies and quality of the live crayfish have dramatically deteriorated. One major problem confronting the crayfish industry is the extended period of time (six months or more) that plants, equipment, and labor may be idled. Some crayfish processors turn to blue crabs (*Callinetes sapidus*) processing in the off season since equipment and labor requirements are similar and sufficient quantities of crabs are available.

SPECIES DIFFERENCES

As previously mentioned, there are two commercially important southern crayfish species, *P. clarkii* and *P. zonangulus*. Harvesting efforts from both managed ponds and wild stocks generally results in a mixture of the two species although *P. clarkii* is by far the predominant species. The species are easily distinguishable by external physical characteristics (Table 16) (Figure 5). In addition, Marshall et al. (1988) evaluated the differences in color, texture, and flavor of processed meat from the two species. Both an instru-

Table 16. A General Comparison of the Red Swamp Crayfish, *Procambarus clarkii,* and the Gulf Coast White River Crayfish, *Procambarus zonangulus,* formerly *Procambarus acutus acutus.*

	P. clarkii	*P. zonangulus*
Color	Basically reddish. Young are greenish-brown. Bases of legs have a red pigment.	Basically whitish. Young are whitish, sandy brown. Bases of legs are devoid of pigment. Young may have many spots.
	Both species are very dark if they come from clear, dark waters or very light and pale if they come from very turbid, murky waters. All become very much darker when mature but *P. clarkii* is much redder while *P. zonangulus* is more purple.	
Carapace	Two halves of carapace almost always join.	Two halves of carapace do not join – an areola is present.
Chelae	Chelae are wide and robust.	Chelae are slender.
Ventral Abdominal Vessel	Blue in color and very apparent.	Translucent and not apparent.
Gonopodium	Club-shaped.	Spear-shaped.
Sperm Receptacle	Distinct S–shape. Poorly indented.	Much more indented.

*Note: Blue, "yellow", and white mutants of *P. clarkii* and blue mutants of *P. zonangulus* have been reported.

mental analysis and a nine-panel sensory panel were used to evaluate samples. Findings showed that sensory panelists could not detect a significant difference in color, texture, or flavor between the two species. These tests were conducted under both red and white light. Even though color differences were more obvious under the white light, the differences were apparently not sufficient enough for the panelists to differentiate.

Instrumental analysis for texture using Instron Kramer shear-force values was in agreement with the panelists, in that no significant differences in texture was observed. However, instrumental evaluations of color using Hunter lab color values showed highly significant differences in the meat from the two species. These differences showed that the *P. clarkii* meat is more red in color and that of *P. zonangulus* is more white in color. In addition, the hepatopancreas (an important edible component) was also evaluated instrumentally for color differences. Again, results showed highly significant differences between the two species. The hepatopancreas of the *P. clarkii* is red-orange in color and changes little over a 20-hr storage period. The hepatopancreas of *P. zonangulus,* however, is greener in color and the green color intensifies over a 20-hr storage period.

Currently, processors do not segregate the two species during processing. Meat and hepatopancreas is packed together in the same proportions of the raw materials received. Note, however, that processors prefer *P. clarkii* because yields are higher. Furthermore, native crayfish consumers note a distinct flavor difference in hepatopancreases of the two species. They generally prefer that of *P. clarkii,* stating that it is "sweeter."

EFFECT OF SOURCE AND HARVESTING

Crayfish harvesting is a highly labor-intensive effort that is generally conducted every day during the season. Quality considerations begin at the time of harvest. Crayfish are harvested in mornings from traps and are delivered to processing plants by afternoon.

Because of the close association of many aquaculture activities with commercial agriculture, concerns about pesticide contamination of edible portions of products often surface. Madden et al.

(1989) conducted a survey of persistent pesticide residues in crayfish abdominal muscle and hepatopancreas, sediments, and growing waters. Samples were collected from aquaculture ponds and wild growing areas in the Atchafalaya River Basin. Results from this study showed that persistent organochlorine pesticide residues in the edible tissue of the commercially important Louisiana crayfish were very low.

Finerty et al. (1990) conducted a survey on metal residue in the edible tissues of crayfish from aquaculture ponds and wild stocks. The toxic metals lead, mercury, and cadmium were not detected from either source.

TRANSPORTING AND RECEIVING LIVE CRAYFISH

Crayfish for processing are received alive from producers. Dead crawfish are not suitable for processing because they decompose rapidly. Crayfish producers and processors must take the necessary steps to minimize environmental stress and to provide conditions that will enhance an extended shelf-life. Under suitable conditions, crayfish can be maintained alive out of water for several days. Harvesters generally pack freshly caught crayfish tightly into mesh bags (Figure 22). These bags are the same type used to pack produce such as onions. Approximately 14-16 kg of crayfish can be packed in each bag. Quickly bagging live crayfish has several advantages (Moody 1989). Tight packing restricts their movement, it minimizes their aggressive behavior, and it facilitates ease in handling and transportation. In addition, the mesh sacks also permit adequate air circulation and permeation needed to keep crayfish alive. Although no published data on factors responsible for extending the shelf-life of live crayfish exist, the industry has shown that crayfish should be stored at cool temperatures, and provided a saturated humidity, and plenty of fresh air. Most processors maintain cooler temperatures around 1.6-4°C. Some processors have found it advantageous to put a thin layer of flaked ice on top of the sacks. Cooling apparently slows down the metabolism of the crayfish and relieves heat stress. A high humidity will slow dehydration. Gentle circulation of fresh air in the storage unit will minimize suffocation.

Harvesters must use caution when handling and transporting crayfish to the processing facility. Even water conditions can have an impact on crayfish condition. For example, crayfish often die quickly when taken from water with low oxygen conditions or when exposed to excessive heat. The natural physiological conditions of the crayfish can also affect crayfish ability to survive. At times, a high percentage of crayfish will simultaneously molt 1-2 days before capture. These individuals have very thin exoskeletons, are extremely sensitive to handling, and often die quickly.

Transporting crayfish from the harvesting site to the processing plant must be done carefully. Sacks of crayfish should not be excessively stacked in order to prevent crushing or restricting the availability of oxygen. This is especially important when handling early season crayfish when the exoskeletons are soft and susceptible to injury. At the same time, sacks of crayfish should not be transported in open bed trucks since the rapid flow of air will quickly dehydrate crayfish gills. Without adequate protection during hot weather, crayfish mortality will increase. Obviously, all contact surfaces from the time of capture to delivery to the processing plant should be free of substances that may contaminate the live crayfish.

When the crayfish are received at the processing plant, each sack is generally individually weighed and tagged with the weight. At this point the crayfish may be processed immediately or stored live in a cooler for later processing or live sales. In order to minimize cross-contamination, these coolers store only live crayfish and are commonly detached from the main facility. Processing plants have other separate coolers for the storage of the finished product.

Live crayfish quality is evaluated when received at the processing plant. This evaluation includes an assessment of the overall condition and the presence of off-odors or foreign material. Poor quality crayfish are rejected or discarded.

WASHING AND GRADING

Grading live crayfish for size has recently become an important quality parameter. Historically, prices paid to harvesters have not been dependent upon crayfish size. Newly developed and significant international markets for premium-sized whole, cooked crayfish have drastically altered domestic marketing traditions. So significant has been the impact of these international markets that today virtually every commercially produced crayfish is graded, whereas only a few years ago grading was not even considered an important part of crayfish processing.

Many international sources of freshwater crayfish have been depleted by disease or overfishing and Louisiana remains a stable supply for many countries where the seafood is considered a tradition. These countries, especially Scandinavia, demand crayfish 30 g or larger. Less than 10% of all crawfish harvested meet this requirement and demand now dictates that significantly higher prices be paid for these large crayfish. Currently, crayfish are generally graded into three sizes: 30 g or larger for the international market, 20-30 g for domestic whole, live crayfish sales and less than 20 g for processing into crayfish tail meat. Grading can occur at the harvesting or processing plant level.

Some harvesters have small static graders on boats that separate marketable crayfish and return small non-marketable crayfish back to the water (Figure 30). Moody (1989) and Gaudé and Gaudé (1989) give an overview of the grading systems used in the industry. Most graders use variably spaced bars or slots to separate crayfish by cephalothorax size, the widest part of the crayfish. The simplest type of grader being utilized by the industry is a series of fixed spaced bars. Crayfish are placed on the bars and the device is shaken by hand or mechanically to facilitate separation. Another type of common grader is a tilted rotating drum made of precisely spaced bars. As the drum rotates, the smaller crayfish fall through the bars and are recovered. The large crayfish are discharged from the open end of the cylinder. This type of grader may have multiple spacing bars for several grades. Other types of graders use small rotating cylinders with graduated spacing (Figures 31 and 32). As the crayfish move down the cylinders from the narrowest spacing to the widest spacing they are graded out. Graders used in the commercial vegetables industry have also had some application for crayfish grading. Modified vegetable graders have also found application in the crayfish industry. Rollers on a belt are engineered so that spacing increases as they pass over the grader. As a result, small crayfish fall out first and largest crayfish fall out last.

Figure 30. A simple on-boat grader using a static grid system to separate two sizes of crayfish. J. Huner.

Figure 31. Grading unit into which crayfish are emptied into the tray on top. Diverging rollers are used on the second tier to separate crayfish into different size groupings. J. Huner.

Crayfish are washed in conjunction with grading or after grading just prior to processing by placing them in large wash tanks filled with clean potable water. These tanks are similar to those used by the shrimp industry for icing shrimp. When placed in the tank the live crayfish settle to the bottom onto a moving conveyor belt. The belt removes the live crayfish from the tank and transports them to

Figure 32. A close-up of the diverging rollers that are part of the grading unit shown in Figure 31. J. Huner.

an inspection line. While in the wash tank, soluble and suspendable materials such as soils and organic materials are removed. Larger foreign materials such as the occasional piece of bait used in the traps and grass are also removed. While on the inspection belt, dead, injured, or defective crawfish can be removed prior to processing. The conveyor belt empties crayfish into the cooking baskets.

At times, a dark-colored scale develops on the underside of the crayfish. It can be especially noticeable on the legs and lower carapace. This scale is considered a defect in whole, cooked crayfish destined for international markets. Attempts to remove this substance have been unsuccessful and further work needs to be conducted in this area. Problems associated with water boatman eggs on the carapace are discussed in the "Disease and Parasite" section that follows.

HEAT PROCESSING

Processed crayfish products generally require heat treatment. The two most common product forms are (1) whole, cooked, and frozen crayfish and (2) cooked, peeled, and deveined crayfish meat.

Proper blanching time is extremely critical for quality crayfish products (Marshall et al. 1987) (Figure 33). In the case of meat production, if crayfish are overcooked, then the manual removal of meat becomes difficult. The intestine (vein) that runs the length of the abdomen (tail) loses the strength necessary for clean removal, requiring extra hand manipulation. A more serious problem occurs if crayfish are undercooked. The adhering hepatopancreas, an important flavor ingredient, is often packaged with ice-stored peeled tail meat. Hepatopancreas that has not been adequately cooked has been implicated in mushiness or softening textural problems in crayfish meat (Marshall et al. 1987). These textural problems are attributed directly to proteolytic enzymes found in the hepatopancreas. With adequate cooking, these enzymes lose activity. A simple in-plant test using gelatin as a test substrate is useful for evaluating enzyme activity and for establishing proper cooking times.

There are no published data that review internal temperatures achieved by crayfish blanching. However, many known factors will affect the proper cooking times for crayfish and include the capacity of the cooking device, the size of the crayfish, the initial temperature of the crayfish, and the water-to-crayfish ratio. Many crayfish processors learn through experience the art of cooking crayfish properly.

Blanching takes place immediately after the crayfish have been washed. There are many varieties of blanchers. However, nearly all of them use boiling water rather than steam to blanch the crayfish. The traditional cooker, and one still common today in processing plants, is simply a metal tank that uses an exposed flame under the tank to heat the water. There are many disadvantages to this meth-

Figure 33. Boiling crayfish in unseasoned water prior to hand-peeling them. J. Huner.

od. Heating is slow, heat transfer is inefficient, and there is some danger with the exposed flame. The biggest advantage is that the system is simple and inexpensive to install. Many modern crayfish processing plants use steam generated from a boiler to heat the water. The steam is either injected directly into the water or into a steam-jacketed kettle. The advantages to steam heating are rapid come-up time, efficient energy use, and clean heating. Although the initial installation cost of steam is considerably higher than gas flame heating, steam heating is much more economical over the long term. In addition, there are other varieties of blanchers. Some use pressurized steam in a retort system to cook the crayfish while some use atmospheric steam to cook the product.

Blanches may be batch types or continuous cookers. When water is used to cook crayfish, the water is kept clean by periodically discarding the cook water after several batches of crayfish. Crayfish intended to be peeled for meat are blanched in unseasoned water. Crayfish blanched for the whole, cooked market are processed in water that may or may not be seasoned.

After blanching, crayfish are air- or water-cooled prior to meat removal (Figure 34) or packaging of the whole, cooked crayfish. Since crayfish meat is traditionally picked warm, many meat producers wait only until the crayfish can be safely handled before proceeding with meat removal. Whole, cooked crayfish are commonly placed in chilled seasoning solution for cooling prior to packaging.

Baskin and Wells (1990) modified a commercial vegetable/fruit steamer to evaluate the use of steam rather than boiling water to prepare crayfish for peeling. A decrease in cooking time and an increase in meat yield were experienced with the system. Physical and microbial characteristics of the steam cooked meat compared favorably to the conventionally boiled product. A computer simulation showed that a 350% throughput increase of crayfish might be achieved compared with conventional boiling methods.

PROXIMATE ANALYSIS

A proximate analysis of *Procambarus* spp. meat washed of extraneous hepatopancreas is provided by Patrick and Moody (1989). Nutrients per 100 g are calories, 89 g; fat, 1 g; protein, 19 g; cholesterol, 139 mg; and sodium, 53 g. The cholesterol content of the hepatopancreas is much higher than that of abdominal muscle, being about 210 mg per 100 g (Reitz et al. 1990). This level is indicative of the high lipid content of this organ.

Figure 34. Hand-peeling crayfish meat ensures high quality but is so costly that automation is necessary if the U.S. industry is to progress. J. Huner.

MEAT PROCESSING

Commercially valuable crayfish meat is removed only from the abdomen, or "tail." Meat is also located in the claws of the crustacean, but no significant markets have been developed for this product; consequently, the claws are discarded as waste (Meyers 1985). As a overall figure, crayfish generally yield approximately 15% meat by weight. Since crayfish grow disproportionately with maturity, there are many variations from this figure. The cephalothorax and claws of mature crayfish are significantly larger. Large mature crayfish may have a meat yield of less than 10% while immature crayfish may have yields higher than 20%. Huner (1988a) compared the meat yield of several categories of crayfish. His findings showed that smaller crayfish, regardless of sex, had highest meat yields. His studies also showed that immature crayfish (male and female) had meat yields that were consistently 4-5% greater than those of mature males. *Procambarus clarkii* produced as much as 3-5% greater yield of edible product than *P. zonangulus.*

Rapidly growing *P. clarkii* have thinner, more flexible exoskeletons than non-molting older ones (Huner et al. 1976, 1988; Silva et al. 1991). There is a change in muscle quality as the crayfish season progresses with both firmness and elasticity increasing to highest levels at the end of the season (Silva et al. 1991). Meat and hepatopancreas yields (%) will decline as the season progresses because the percentage of low-yield, mature males in the population increases (Huner 1978b, 1988a).

Meat yield can also be affected by other factors (Marshall and Moody 1986) such as cooking times and meat removal practices. Meat removal is done manually by a cadre of workers. Crayfish are individually peeled of shell and deveined. The procedure is time consuming and adds considerably to the production cost. Generally workers are paid by the number of pounds of meat removed.

Although there have been many attempts to mechanize meat removal, the industry has not adopted any of the technology (Huner and Barr 1991). Moody (1989) gives a summary of the mechanical devices that have been devised and tested in the industry. The two principle methods are rollers that squeeze the meat from the abdomen or compressed air delivered by a needle to push the meat from the abdomen. "Deheading" has been a problem and difficulty has been experienced in developing automated deheading devices.

The largest room of a crayfish processing plant is the meat peeling room. In this room, workers are stationed at long, narrow stainless steel tables. Freshly blanched crayfish are provided to the workers by piling them in the middle of the table. The skilled workers deftly remove the meat and discard the waste. The meat is placed into sanitary colanders and the waste placed into marked containers removed from the table. After each pile of crayfish has been peeled of meat, the table is sanitized and the procedure starts all over again. When the colander is full of freshly peeled meat, it is delivered to the packaging room for packaging. Colanders are generally emptied frequently in order to maintain quality.

The procedure for peeling crayfish requires that the meat be removed in one solid piece along with the naturally adhering hepatopancreas. The hepatopancreas, referred to as "fat" by the industry and consumers, is an extremely important finished crayfish meat constituent. It adds flavor to crayfish dishes and its bright orange color adds eye appeal. It is considered an edible portion of the meat and consequently the naturally adhering "fat" is included in the net weight of the product. Marshall and Moody (1986) evaluated the factors that contribute to hepatopancreas yield on the final product. In addition to blanch time and peeling practices, the time of year also had an effect on hepatopancreas yield. As a general trend, the percentage of hepatopancreas yield increased with the progression of the wild crayfish season. Throughout the fishing season, the hepatopancreas tissue yield varied from 2.2 to 13.1% of the final product. The season average was 8.14%.

PACKAGING AND STORAGE

After delivery to the packaging room, the meat is spread in trays and examined for extraneous material such as remaining pieces of intestines (veins) and bits of shell. Meat to be sold as fresh product is generally placed in polyethylene bags, weighed, air compressed out of the bags by hand and heat sealed for closure (Figures 35 and 36). Immediately after packaging, the bagged meat is submerged into an ice-water slush for quick chilling. Fresh meat is always packed with the natural adhering hepatopancreas. The shelf-life of the fresh meat, like most seafoods, is dependent upon many factors. Generally, the typical shelf-life is from 7 to 12 days held under ideal conditions.

Meat to be frozen is packed with or without the natural adhering

Figure 35. Weighing and packing crayfish meat. J. Huner.

hepatopancreas. Commercial experience has shown that frozen crayfish meat packed with hepatopancreas and packaged in polyethylene bags may develop off-flavors and rancidity after two months of frozen storage. The use of vacuum packaging and laminated bags designed to prevent dehydration and oxygen permeation greatly retard the onset of rancidity. Many processors remove hepa-

Figure 36. Packed crayfish meat. Basic unit is 454 grams (1 pound). J. Huner.

topancreas tissue by washing in cool, clean water, because the material is water soluble and is easily washed away with little agitation.

Marshall (1988) did an extensive study on frozen crayfish meat. Findings from this study clearly show that texture was more dependent upon frozen storage time rather than by the freezing methods evaluated. This study also showed that meat frozen by cryogenic freezing techniques had significantly less drip loss associated with the final product than meat frozen using more conventional methods. It is also interesting to note that meat frozen with the hepatopancreas had a significantly higher drip loss than washed meat. Cryogenically frozen meat also had a more desirable texture than conventionally frozen meat.

Frozen crayfish meat can sometimes develop a dark blue discoloration when thawed and reheated. This discoloration is a harmless phenomenon, but unappetizing. It seems to occur more often with late season crayfish. Research by Moody and Moertle (1986) showed that treatment with a chelating agent prior to freezing is effective in minimizing discoloration. The work showed that both citric acid and EDTA demonstrated varying degrees of effectiveness in preventing discoloration. A 1.25% citric acid dip is recommended and approved for minimizing this problem.

WHOLE, COOKED, AND FROZEN CRAYFISH PROCESSING

Whole, cooked, and frozen crayfish have become increasingly important to commercial crayfish processors. There is considerable demand for the product in European markets, especially Sweden. In addition, processors are also finding increased local demand for this traditional dish. This product requires large crayfish free of defects such as missing claws and legs. After blanching, the whole crayfish are cooled in seasoned or unseasoned water or steamed and packaged in trays. Additional seasoned water is packed with the crayfish to eliminate air pockets and protect the product against dehydration (freezer burn) and oxidation. Most modern plants that are producing this product use cryogenic freezing systems to freeze the packaged product prior to cold storage. Cole and Kilgen (1987) studied factors related to the product shelf-life. Their studies showed that whole, cooked crayfish that were individually frozen using cryogenic freezing systems were superior to other types of processing.

PURGING

Purging should not be confused with the traditional use of 400-500 grams of salt placed in a large tub of water with 15-20 kg of crayfish before they are boiled. Although this is referred to as "purging," it only causes the crayfish to regurgitate their stomach contents. It does not clear the gut, a process that takes 1-2 days depending on temperature. External organisms such as branchiobdellid worms are also killed and drop from the crayfish during the salt bath process.

True purging of crayfish involves holding crayfish for 1-2 days

Figure 37. A cage used to hold and purge crayfish before sales. J. Huner.

until their guts are naturally emptied. Lawson and his students (Lawson and Baskin 1985; Lalla and Lawson 1987; Lawson et al. 1990) describe research efforts to develop cost-effective purging systems. Two kinds may be used: the submerged system and the spray system. In the submerged purge system, crayfish are completely submerged in water which may be flow-through or static with periodic total or partial changes (Figure 37). In the spray purge system, crayfish are held in the bottom of an empty container and small amounts of water are sprayed on them periodically. This reduces the amount of water required for purging and simplifies waste control problems. Details on recommended flow and loading rates for the two types of purging systems are provided in Table 17.

A considerable amount of silt and clay can adhere to crayfish. Submersion for 1-2 hours will eliminate this. This product will cook better and there will be no "muddy" flavor. However, the digestive tract will not be "purged" of fecal material. Many restaurants are using this "purging" method in lieu of a true purge because it is less expensive and consumers react positively to the product.

Crayfish processors report that purged crayfish are much more robust than unpurged crayfish. Certainly, there are losses during the purging process because weak crayfish die and there is weight loss from the actual purging of the gut contents. These costs of operation must be recovered if a purging system is to be profitable.

WASTE UTILIZATION

A major obstacle for crayfish processors is the large amount of waste generated resulting from meat processing (Figure 38). Utilizable tail meat yields are typically about 15% leaving 85% waste. This waste is composed primarily of the entire exoskeleton, internal organs from the cephalothorax, and meat from claws and legs. Louisiana crayfish processing plants produce an estimated 40+ million kg

Table 17. Comparison of Crayfish Purging Systems, After Lawson and Baskin (1985), Lalla and Lawson (1987), Lawson et al. (1990).

System	Loading Rate	Water Flow Rate	Survival-40 Hrs
Submerged			
Batch	—*	Complete Exchange 3-4 times/day	—*
Flow-Through	24.5 kg/m^2	60.6 liter/min.	85-90%
Spray	24.5 kg/m^2	3.6 liter/min.	90%

*No comparison data available. Loading rate is assumed to be lower than other systems with lower survival.

Figure 38. Crayfish wastes carried from a processing room by a conveyer belt. Wastes account for about 85% of the total weight of the crayfish. J. Huner.

of waste annually (Tanchotikul and Hsieh 1989). Typically, waste is disposed of in land fills or by other similar methods. Growing environmental concern will require alternate means of disposal or utilization. Researchers have evaluated by-product utilization of crayfish.

Pigment extraction from the exoskeleton of crayfish has been shown to be a commercially feasible process when used in aquaculture dietary formulations. These pigments impart desirable coloration to edible flesh portions. Astaxanthin recovered from crayfish waste is currently being used in conjunction with red sea bream mariculture in Japan (Meyers 1987). Meyers and Bligh (1981) characterized the astaxanthin pigments present in the heat-processed exoskeleton of crayfish. Results of this study demonstrated a pigment concentration as much as 20 times greater than that reported for commercial shrimp meals. Extracted pigments were astaxanthin ester (49.4%), astaxanthin (40.3%), and astacene (10.3%). Pigment concentrations were as great as 153 µg/g. Chen and Meyers (1983) demonstrated a 40 to 50% increase in extracted astaxanthin pigment concentration by using acid ensilage treatment.

Chitin recovery (23.5% on a dry basis) from crayfish waste has been shown to be feasible (No et al. 1989). Chitosan derived from crayfish exoskeleton chitin has been shown to effectively coagulate suspended solids in seafood processing waste water (No and Meyers 1989a) and to effectively recover amino acids from seafood processing waste water (No and Meyers 1989b).

Tanchotikul and Hsieh (1989) demonstrated the presence of important crayfish flavor components in crawfish processing waste. Many of the same flavor compounds identified in crayfish tailmeat were also detected in the waste. Recovery of these flavor components may eventually prove to be commercially and economically feasible.

Fresh and dried crayfish processing wastes do have agricultural value. They dry in a short time, even in humid climates when spread thinly over fields. Desselle (1980) studied the usefulness of crayfish wastes as a lime source and fertilizer for pastures and presented recommendations for application rates. Barry (1980) studied the usefulness of raw crayfish wastes in production of vegetable cropsand also made recommendations for application rates. Wastes may also be composted with substances such as rice hulls to be used as plant potting media. The two major problems involved in use of raw crayfish wastes in crop situations are the cost of transporting the wet material and the potential for pollution of surface and ground waters.

SOFT-SHELL CRAYFISH PRODUCTS

Soft-shell crayfish may be sold alive or frozen (Figure 39). The live crayfish are held in water chilled to 4-6°C and survive up to 7 days. The stomach stones, or gastroliths, should be removed before soft-shell crayfish are cooked. These paired, hemispherical calcium carbonate structures are formed in the lining on opposite sides of the stomach just behind the eyestalks. They are "molted" into the lumen of the stomach at the molt. A vertical cut just behind the eyes exposes them and they are readily removed with minimal pressure.

The hepatopancreas and/or the carapace may be removed especially if the cleaned product is to be "stuffed." Alternatively, the body may be split between the legs to make a "butterfly" presentation that is suitable for stuffing.

Soft-shell crayfish are usually fried or broiled. A number of tasty recipes have been developed or adapted from various soft-shell blue crab recipes. These all make very attractive table presentations when properly garnished (Figure 40).

Figure 39. Soft-shell crayfish are frozen alive in trays for storage. J. Huner.

Figure 40. Soft-shell crayfish can be prepared according to a number of tasty recipes and eaten whole as long as the gastroliths in the anterior part of the cephalothorax are removed. J. Huner.

Section III:
Diseases of Louisiana Crayfish

INTRODUCTION

Diseases of crayfish encompass a range of problems with different levels of significance. Scientific literature addresses primarily metazoan parasites, only a couple of which may have an impact on commercial crayfish production. Crayfish culture as currently practiced in the United States is relatively free of significant disease problems. Only recently, with the development of high density systems for purging and for soft-shell production, have conditions similar to those experienced in more intensive aquaculture systems been encountered. This intensification has led to the emergence of some significant disease problems. Additionally, there are some agents that do not limit production, but may affect marketing due to their effects on the appearance of the product.

BACTERIA

Potential bacterial pathogens of crayfish include species of the genera *Aeromonas, Bacillus, Corynebacterium, Mycobacterium, Flavobacterium, Pseudomonas, Vibrio, Edwardsiella*, and *Acinetobacter*. Members of these genera are common inhabitants of the stagnant water and decomposing vegetation found in crayfish production ponds and have been isolated from the crayfish exoskeleton, hindgut, and hemolymph. They are also associated with opportunistic bacterial infections in other aquatic species and have been implicated in external and internal infections in cultured crustaceans.

The chitinous cuticle and lipid containing epicuticle provide crayfish with an effective barrier against the invasion of most micro-organisms. However, chitinoclastic bacteria can cause a syndrome in crayfish and other crustaceans known as shell disease, which is characterized by black or dark brown lesions on the exo-skeleton (Figure 41). These lesions are due to melanization of a wound or damaged area of the exoskeleton associated with various bacteria, primarily of the genera *Aeromonas, Pseudomonas, Acinetobacter,* and *Vibrio.* The incidence and severity of this syndrome can increase under conditions where molting is infrequent (Figure 42), as in hold-ing facilities or in nutritionally poor environments. Intensive systems also favor the development of shell disease, probably due to the infection of wounds caused by aggressive behavior, increased ease of pathogen transfer due to increased density, and poor water quality.

Systemic bacterial infections are generally considered to be rare in crayfish in commercial ponds. However, evidence indicates that significant levels of bacteria can occur in the hemolymph of pond raised crayfish (Scott and Thune 1986a) and that these increases are

Figure 41. Dark lesions on the lateral surface of a crayfish with bacterial shell disease. R. Thune.

Figure 42. Severe case of bacterial shell disease demonstrating a lesion that has penetrated the exoskeleton to the gill chamber. R. Thune.

dependent on high temperature and low dissolved oxygen levels. Stress associated with extremes of either of these factors seems to affect the ability of crayfish to control bacterial populations in the hemolymph. These bacteremias are atypical of those encountered in other cultured aquatic species in that clinical signs of disease are not apparent and that multiple bacterial genera are present in an individual, rather than one or two obligate or opportunistic pathogens.

A recent report (Thune et al. 1991) documents the first cases of bacterial septicemia in crayfish. *Vibrio mimicus* and *V. cholera* were the predominant organisms isolated in each of 15 cases examined. Of the 15 cases, 1 occurred in a holding tank, 4 in ponds, 5 in purging systems and 6 in soft-shell operations (Table 18). Epizootics occurred near the end of the pond harvest season in late May and early June, during periods of elevated water temperatures and/or depressed levels of dissolved oxygen. Diseased crayfish were lethargic and were easily identified by their failure to posture when

Table 18. Summary of conditions for vibriosis cases in crayfish in Louisiana for 1985-1990.

Case	Organism	Temp. (C)	System
85-63	*V. cholerae*	26	Purging
85-69	*V. mimicus*	26	Purging
86-56	*V. mimicus/V. cholerae*	23	Pond
86-77	*V. mimicus*	23	Purging
86-110a	*V. mimicus*	30	Purging
86-111	*V. mimicus*	30	Purging
86-120	*V. mimicus*	30	Pond
87-123	*V. mimicus*	30	Soft-Shell System
87-136	*V. mimicus*	26	Pond
87-240	*V. mimicus/V. cholera*	26	Soft-Shell System
88-13	*V. mimicus*	26	Soft-Shell System
89-255	*V. mimicus*	28	Ponds
89-323	*V. mimicus*	28	Holding tank
90-297	*V. mimicus*	25	Soft-Shell System
90-327	*V. mimicus*	25	Soft-Shell System

From Thune et al. (1991)

threatened, yet they exhibited no gross clinical signs. Daily mortality rates range from 5 to 25%. Physiological, biochemical, and serological profiles for *V. cholera* and *V. mimicus* isolates indicated each crayfish isolate was virtually identical in biochemical profile with the exception of sucrose fermentation, which differentiates these two vibrio species (Davis et al. 1981). Serological evaluation indicated that each isolate expressed a single 0 antigen, serovar J, using the system and nomenclature of Adams and Siebeling (1984). Serovar J corresponds to Sakazaki serovar 6 (Sakazaki and Shimada 1977) and Smith serotype 14 (Smith 1979).

The predominant *Vibrio* species known to produce septicemias in other crustaceans are *V. parahemolyticus*, *V. alginolyticus*, and *V. anguillarum*. Since crustaceans previously reported to be infected by *Vibrio* sp. are marine species, it is not surprising that halophilic *Vibrio* species have been the predominant species reported. According to Baumann et al. (1980) optimal NaCl concentrations for *V. anguillarum* are 60-100 mM, 160-180 mM for *V. parahemolyticus* and 200-260 mM for *V. alginolyticus*. *Vibrio cholera* requires an optimal range of only 5-15 mM NaCl, which may account for its association with the freshwater crayfish.

The typability of *V. mimicus* strains with *V. cholera* antisera is not unusual (Davis et al. 1981); however, the isolation of a single *V. cholera* non-01 serotype from affected crayfish from several parishes across south Louisiana over a period of several years is surprising because non-01 vibrios are serologically diverse (Smith and Goodner 1965; Sakazaki and Shimada 1977; Smith 1979; Adams and Siebeling 1984). That a single serotype was responsible for all reported cases indicates a common functional characteristic associated with the bacterial cell wall that favors the establishment of infections in crayfish. Additionally, the serovar in question may have public health significance. All 46 isolates of serovar 6 evaluated by Sakazaki and Shimada (1977) were of human origin, although the clinical status of the patients was not disclosed. Twenty-six of 107 isolates of serovar 14 evaluated by Smith (1979) were associated with gastrointestinal disease and 2 others with non-gastrointestinal illness. However, no documented case of human vibriosis has been correlated to crayfish consumption.

Although the overall impact of both the clinically inapparent bacteremias and vibriosis on the crayfish industry is unknown, high levels of bacteria associated with water temperatures greater than 27°C, low dissolved oxygen levels, transport from ponds to intensive systems, or a combination of these factors may effect crayfish production. It is possible that ponds drained late in the season, when the water temperatures are warm, will have reduced survival of crayfish in the burrows. This would lead to a loss of "brood" crayfish, potentially reduce recruitment of young the following year, and also reduce the crop of holdover crayfish the following fall. Additionally, increased bacterial levels may affect survival and quality of crayfish during transportation, storage, and handling after harvest, especially when they are placed in intensive systems.

FUNGI

Crayfish plaque fungus, caused by the fungi *Aphanomyces astaci*, is notorious for the elimination of native crayfish populations in many European waters and has been extensively studied in European crayfishes of the genera *Astacus* and *Austropotamobius*. The fungus attacks almost entirely in the soft cuticle in joints or between segments and can kill a susceptible crayfish in a week or two. North American crayfish appear to be resistant to severe pathological affects of *A. astaci,* although limited, chronic infections can be established (Unestam 1969b). Recently, the relationship between the North American crayfish and the crayfish plaque fungus has come into question due to the increased exportation of live crayfish to Europe. Both *Pacifastacus leniusculus* and *Orconectes limosus* are considered to be vectors of *A. astaci* in Europe (Alderman and Polglase 1988; Vey et al. 1983). Furthermore, *Procambarus clarkii* from Spain have been shown to be potential vectors in Swedish laboratory studies (Uribeondo and Söderhäll 1992).

A related species, *Aphanomyces laevis*, has been described as primarily a wound parasite for red swamp crayfish, but can cause mortality in crowded, unhealthful conditions (Smith 1940). It differs from *A. astaci* in having narrower hyphae, smaller oogonia, and a thin smooth oogonial wall rather than a thickened rough one. *A. laevis* is generally considered to be more saprophytic than parasitic (Coker 1923) and could not reliably induce disease in inoculated crayfish (Smith 1940; Unestam 1969a).

Other fungi known to invade the cuticle and flesh of crayfish include the species of the genera *Fusarium* and *Ramularia*. *Fusarium* spp. can cause significant disease in marine shrimp and lobsters, although mortalities associated with this organism have not been described in crayfish. *Ramularia* causes "burn spot disease," which appears as round brown or black spots in the exoskeleton, 1 cm or more in diameter, and often with raised margins (Unestam 1973) (Figure 43). The development of both of these infections in crayfish is favored by crowded conditions and breaks in the exoskeleton.

Psorospermium haeckeli is a unicellular organism of uncertain taxonomic status that is commonly seen in the tissues of Louisiana crayfish (Hentonnen et al. 1991). It has variously been classified as a protozoan (Grabda 1934; Schaperclaus 1954; Unestam 1975) and a nematode or trematode egg (Ljunberg and Monné 1968; Schaperclaus 1979), but current thought is that it is a fungal organism of some type, possibly a species of the dimorphic pathogenic fungi (Nylund et al. 1983; Alderman and Polglase 1988). The organism is detected microscopically as a distinctive ovoid object with a thickened cell wall and a highly vacuolated cytoplasm (Figure 44); however, thin cell wall forms are encountered regularly. There appears to be little or no host response to its presence in the tissue and no reports have been made of any mortality associated with the presence of this organism.

A fungal group known as Trichomycetes are common in crayfish

Figure 43. *Ramularia,* or "burn spot," on the dorsal aspect of a red swamp crayfish. R. Thune.

Figure 44. Characteristic form of *Psorospermium haeckeli* in crayfish tissue wet mounts. R. Thune.

intestines and occasionally on the cuticle. However there is some question as to whether these organisms should be considered parasites or commensals (Johnson 1977).

PROTOZOANS

Protozoa associate with crayfish as true parasites and as ecto-commensals. Genera of the order Microsporida are the principal parasitic forms, while ectocommensals are genera of the subphylum Ciliophora.

The microsporida are intracellular parasites of invertebrates and lower vertebrates, with some species being quite pathogenic. Microsporideans infect crayfish when microscopic spores ingested by the host extrude a filament that deposits an infective stage through the gut wall. This infective stage presumably migrates to specific tissues to invade a host cell, where it divides repeatedly, producing large numbers of spores. In crayfish the spores develop primarily in muscle and cause the tissue to take on a milky-white appearance (Figure 45), especially when substantial muscle tissue has been affected. In an advanced infection, animals are noticeably sluggish with an ineffectual tail-flick response. Histologically, normal muscle tissue is replaced by microsporidian cysts. Although little

Figure 45. Crawfish exhibiting the milky-white appearance of the abdominal muscle associated with a severe infestation of the microsporidean *Thelohania*. Normal crayfish on the right. R. Thune.

pathology or host response is evident, an infiltration of eosinophilic granular cells is apparent (Figure 46).

Microsporidiosis in crayfish, commonly known as porcelain disease in Europe, is primarily due to the genus *Thelohania* in the class microsporididea. Other genera of this class such as *Pleistophora*, *Nosema*, and *Inodosporaes* have been described from other crustaceans such as marine shrimp. *Thelohania* has microscopic uninucleate spores (Figure 47) produced by synchronous sporogony resulting in pansporoblasts containing 8 spores.

Ectocommensal protozoans of the phylum *Ciliophora* utilize the crayfish cuticle as a substrate for attachment and rely on bacterial populations in the water for nourishment (Fisher 1977; Couch 1983). Accumulations of suspended organic material often occur in commercial crayfish ponds in the spring, as indicated by increased turbidity. When this occurs, bacterial concentrations also increase, allowing ectocommensals to flourish (Figure 48). A study by Scott

Figure 46. Micrograph of a section of crayfish abdominal flexor muscle infested with the microsporidean *Thelohania*. Note the general lack of pathology and the infiltration of the tissue with large eosinophilic granulocytes (1). R. Thune.

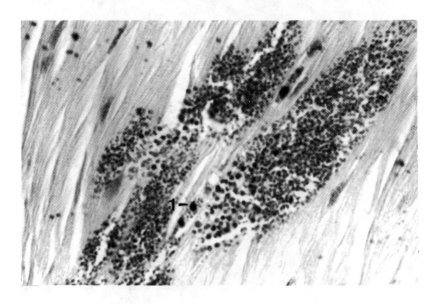

Figure 47. Phase contrast micrograph of *Thelohania* spores from the abdominal flexor muscle of an infested crayfish. R. Thune.

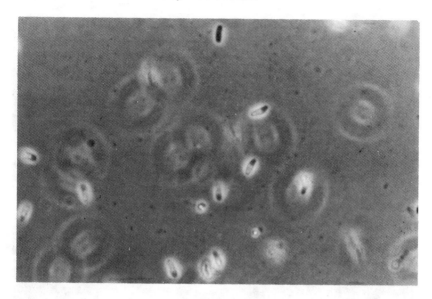

and Thune (1986b) indicated that 94% of crayfish sampled from commercial ponds were infested, with 65% of the infested crayfish carrying more than 100 ectocommensals per gill filament.

Ectocommensals most commonly associated with crayfish (Figure 49, A-D) are the peritrichs *Cothurnia, Epistylis, Lagenophrys,* and *Zoothamnium* and the suctorian *Acineta,* with the presence of any particular group depending on water quality. The peritrichs feed on bacteria and are favored by high bacterial levels in the water column (Sleigh 1973; Sawyer et al. 1979), while *Acineta* feeds primarily on free-swimming ciliated protozoans (Hall 1979; Sawyer et al. 1979). Turbidity is an excellent water quality indicator of potential peritrich infestations in commercial crayfish ponds (Scott and Thune 1986b), particularly for *Cothurnia* and *Epistylis,* the most commonly found genera. *Lagenophrys* is generally observed in low numbers, primarily in the spring. *Zoothamnium* is not common in the gill chamber of crayfish, but is often found on the exoskeleton.

Figure 48. Scanning electron micrograph of crayfish gill filaments carrying a heavy infestation of the ectocommensal protozoan *Cothurnia*. R. Thune.

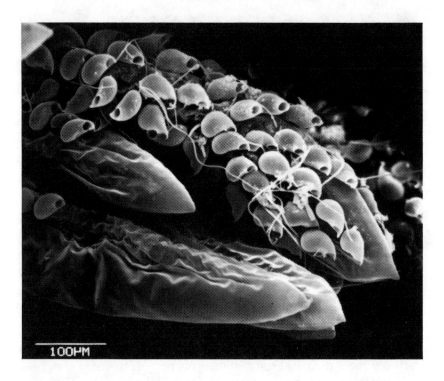

Ectocommensal protozoans do not damage the gill surface (Vogelbein and Thune 1988), but their presence may result in increased susceptibility to low dissolved oxygen levels due to decreased gill surface area and impaired water flow through the gill chamber. Although several authors (Johnson 1977; Huner and Barr 1991) have postulated that ectocommensal protozoa can cause mortalities in commercial crayfish ponds, this has never been demonstrated with certainty. During periods when environmental conditions are favorable, crayfish can tolerate heavy branchial infestations without apparent harm and the infestations can spread to the exoskeleton, occasionally in high numbers without mortality (Figure 50). Whether or not these infestations increase the susceptibility of crayfish to the poor water quality conditions that periodically occur in com-

Figure 49. Ectocommensal protozoans commonly associated with Louisiana crayfish. A. *Epistylus,* B. *Acineta,* C. *Lagenophrys,* D, *Cothurnia.* R. Thune.

Figure 49 (continued)

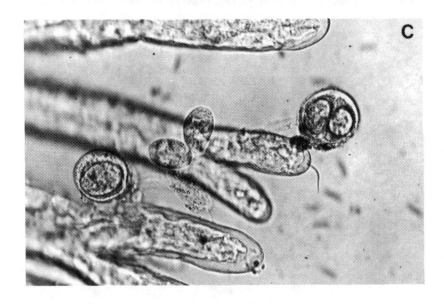

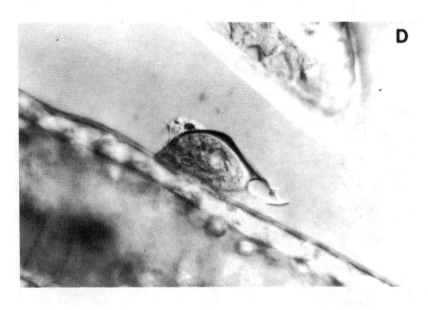

Figure 50. Extremely heavy infestation of *Epistylus* on red swamp crayfish from a pond with depleted forage during early winter. Numerous animals collected from this pond had loads ranging from a few patches to that pictured. R. Thune.

mercial ponds requires further study. However, critically low dissolved oxygen levels are most prevalent in the fall when ectocommensal levels are low due to lower food availability for the protozoans, and to the fact that young, rapidly growing crayfish molt frequently, which reduces their protozoan load. Thus, if ectocommensal protozoans limit commercial crayfish production in ponds, the effect would occur during occasional low dissolved oxygen levels in the spring, when suspended nutrient levels increase and mature crayfish that molt infrequently are predominant. Some mortalities of crayfish in soft-shell production systems have been associated with heavy infestations of protozoans on the gills. These crayfish exhibited a "wooly" appearance and died in molt (Huner unpublished).

One other group of ciliates found on crayfish belongs to the subclass apostomatia and is characterized by having a complex life cycle. One stage, the phoront, encysts on the crayfish cuticle. When the host molts, it exencysts and becomes a trophont which feeds on exuvial fluids or tissues of the dead or molted host. After a series of mitotic divisions, free-swimming tomites that can attach to new hosts are produced (Lee et al. 1985). Common genera in North American fresh and brackish water are *Hyalophysa*, *Gymnodinioides*, and *Terebrospira*, with *Hyalophysa* being more common in freshwater (Johnson 1977).

PARASITIC WORMS

Digenetic Trematodes

Only two groups of parasitic worms are described in crayfish. The first of these, members of the phylum Platyhelminthes, subclass Digenea, are characterized by having a complex life cycle involving two or more hosts. Most worms that parasitize crayfish are present as juvenile forms known as metacercaria, which develop into adult worms when ingested by the appropriate final host. Table 19 lists the most common metacercaria of Louisiana crayfish, their location in the tissues, and other hosts involved in the life cycle.

One particular digenea, *Paragonimus kellicotti*, deserves special mention because the adult develops in the lung of crustacean-eating mammals, including cats and dogs. Clinical signs, including dullness and intermittent cough, are mild. Human infections, although rare, have been reported. *Procambarus clarkii* also serves as an intermediate host for *P. westermani,* a potential human pathogen, in Japan (Hamajima et al. 1976). Thorough cooking eliminates the possibility of human infestation with either parasite. The metacercaria are found encysted in the heart of naturally infected crayfish.

Two digenetic trematodes, *Allocorrigia filliformes* and *Alloglossoides cardicola*, that utilize the crayfish as the final host and reside in the antennal gland have been described. Turner (1984) studied *A. filiformes* in *P. clarkii* and determined that the worm resides with its anterior end in the nephridial tubule of the antennal gland and its posterior end often extending into the excretory bladder. No host response was evident to the worm itself, but melanized nodules

Table 19. Common Metacercaria for Louisiana Crayfish.

	Location	First Intermediate Host	Final Host
Crepidostomum cornutum	Hepatopancreas, heart, cephalothorax	Sphaeriid clam, *Musculium*	Fish
Gorgodera amplicava	Stomach wall		Amphibians
Microphallus opacus	Hepatopancreas	*Amnicola*	Various vertebrates
Maritrema obstipum	Gill filaments and Hepatopancreas	*Amnicola*	
Macroderoides typicus	Cephalothoracic musculative	*Heliosoma*	*Amia calva*
Paragonimus kellicotti	Heart and surrounding membranes		Cats, mink, skunks, raccoons, fox, and other crawfish-eating mammals
Ochetosoma	Abdominal muscle	Physid snails	Watersnakes

containing trematode ova were observed within the interstices of the nephridial tubule. Turner (1985) describes *A. caridicola* as residing in the lumen of the nephridial tubules, where minute tegumental spines abraded the tubule epithelium. However, no host response to either worm was observed.

Spiny-headed worms

The second group of parasitic worms found in crayfish are the spiny-headed worms, represented by a single species, *Southwellinia dimorpha* (phylum Acanthocephala). *Southwellinia dimorpha* uses the crayfish as an intermediate host, encysting in the anterior portion of the abdomen, usually in association with the intestine. Adult worms of *S. dimorpha* are found in the small intestine of the white ibis, *Eudocimus albus* (Schmidt 1973). An embryonated egg containing the infective acanthor stage is released in the feces. Development proceeds in the intermediate crayfish host when the egg is eaten and the acanthor is released to penetrate the gut wall and encyst in the tissues surrounding the intestine. The worm then becomes parasitic, absorbing nutrients and developing into a rice-grain-sized white to pink cyst (Figure 51) containing the infective stage, called a cystocanth. The intermediate host must then be eaten by the final host where the adult develops and the life cycle is completed.

Southwellinia dimorpha seems to be more prevalent in ponds that are not completely dewatered and dried during the summer. If heavy infestations occur, market acceptability can be adversely affected due to the unsightly appearance when the tail is removed. In

Figure 51. Rice-grain-sized cystocanth cyst of *Southwellinia dimorpha* (1) along the intestine of a red swamp crayfish just posterior to the cephalothorax/abdomen junction. R. Thune.

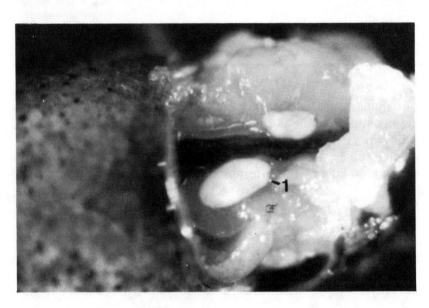

addition, some cysts tend to remain attached to the tail meat when peeled, necessitating additional processing to remove them prior to packaging in commercial peeling operations.

MISCELLANEOUS

Members of the order Branchiobdellida in the phylum Annelida have an anatomy similar to oligochaete worms and leeches. Holt (1973, 1975) separates them from oligochaetes by the lack of setae, the fixed number of segments, the presence of suckers and the unpaired gonopores. He differentiates them from leeches by the anterior position of the testes and by the different number of segments. They are generally ectocommensal on the crayfish carapace, but some species are parasitic in the gill chamber of the host (Holt 1973, 1975). There are several reports of branchiobdellids found away from the host (Young 1966; Bishop 1968; Holt 1973), but little is known of their life cycle. It is presumed that they are dispersed primarily by body contact and that infestations are increased and maintained on a given host by the hatching of eggs commonly found on the carapace (Huner and Barr 1991).

A variety of other invertebrates are found inhabiting the surface of the crayfish carapace. Certain hemipteran insects known as water boatmen lay their eggs on the shell of crayfish and can become numerous when populations of these insects are high and crayfish molting frequency is low. A variety of ostracod crustaceans, rotifers, temnocephalid flatworms, leeches, and nematodes are also found as free-living inhabitants of the crayfish cuticle, but, as with the other ectocommensal organisms described previously, they cause little damage and are purged during molting.

Uropod swelling is commonly observed in crayfish but has no apparent adverse effects. This syndrome seems to be more apparent in the spring when *Procambarus* spp. are growing rapidly (Huner and Barr 1991).

Older mature *Procambarus* spp. may undergo severe muscular atrophy followed by death, presumably associated with old age. Lindqvist and Mikkola (1979) first described this syndrome histologically in a *P. clarkii* population in Kenya.

LITERATURE CITED

Adams, L. B., and R. J. Siebeling. 1984. Production of *Vibrio cholerae* 01 and non-01 typing sera in rabbits immunized with polysaccharide-protein carrier conjugates. *Journal Clinical Microbiology* 19:181-186.

Aiken, D. E. 1969. Ovarian maturation and egg laying in the crayfish *Orconectes virilis*: influence of temperature and photoperiod. *Canadian Journal of Zoology* 47:931-935.

Aiken, D. E., and S. L. Waddy. 1987. Molting and growth in crayfish: a review. *Canadian Technical Report of Fisheries and Aquatic Sciences* 1587:1-34.

Alderman, D. J., and J. L. Polglase. 1988. Pathogens, parasites and commensals. Pages 167-212 in D. M. Holdich and R. S. Lowery (eds.), *Freshwater Crayfish–Biology, Management and Exploitation*. Croom Helm, London, England & Sydney, Australia and Timber Press, Portland, Oregon, USA.

Ameyaw-Akumfi, C. 1981. Courtship behavior in the crayfish *Procambarus clarkii* (Girard) (Decapoda, Astacidae). *Crustaceana* 40:52-64.

Andrews, E. A. 1906. Egg-laying of crayfish. *American Naturalist* 40:343-356.

Andrews, E. A. 1907. The young of the crayfishes, *Astacus* and *Cambarus*. *Smithsonian Contributions to Knowledge* 35:5-79.

Austin, C. M. 1991. Effect of temperature and salinity on the survival and growth of juvenile red claw (*Cherax quadricarinatus*). *Program and Abstracts, 22nd Annual Conference & Exposition, World Aquaculture Society*, San Juan, Puerto Rico. p. 16.

Avault, J. W., Jr., and M. W. Brunson. 1990. Crawfish forage and feeding systems. *Reviews in Aquatic Sciences* 3:1-10.

Avault, J. W., Jr., and A. E. Granados. 1990. Economic implications of pond rotation of the prawn *Macrobrachium rosenbergii* with the red swamp crawfish, *Procambarus clarkii*. *Program and Abstracts of the 8th International Symposium of Astacology*, Louisiana State University Agricultural Center, Baton Rouge, Louisiana, USA, p. 26.

Avault, J. W., Jr., and J. V. Huner. 1985. Crawfish culture in the United States. Pages 1-62 in J. V. Huner and E. E. Brown (eds.), *Crustacean and Mollusk Aquaculture in the United States*. AVI Publishing Co., Westport, Connecticut, USA.

Avault, J. W., Jr., L. W. de la Bretonne, Jr., and J. V. Huner. 1975. Two major problems in culture of crawfish in ponds: oxygen depletion and overcrowding. *Freshwater Crayfish* 2:139-144.

Avault, J. W., Jr., R. P. Romaire, and M. R. Miltner. 1983. Red swamp crayfish, *Procambarus clarkii*, 15 years research at Louisiana State University. *Freshwater Crayfish* 5:362-369.

Baker, F. 1987. Pumps and pumping efficiency. Pages 70-73 in R. Reigh (ed.), *Proceedings 1st Louisiana Aquaculture Conference*, Louisiana Cooperative Extension Service, Louisiana State University, Baton Rouge, Louisiana, USA.

Bankston, D. J., F. E. Baker, T. Lawson, and J. Roux. 1989. Demonstration of paddlewheel aerators in crawfish ponds. *American Society of Agricultural Engineers/Canadian Society of Agricultural Engineers* Paper No. 897012, 7 pp.

Barr, J. E., J. V. Huner, D. P. Klarberg, and J. Witzig. 1978. The large invertebrate-small vertebrate fauna of several south Louisiana crawfish ponds with emphasis on predaceous arthropods. *Proceedings of the World Mariculture Society* 9:683-692.

Barry, R. 1980. Utilization of crawfish peeling plant waste as soil amendment for vegetable crop production. Pages 85-91 in D. Gooch and J. V. Huner (eds.), *Proceedings First National Crawfish Culture Workshop*. University of Southwestern Louisiana, Lafayette, Louisiana.

Baskin, G. R., and J. H. Wells. 1990. Evaluation of alternative cooking schemes for crawfish processing. *Journal of Shellfish Research* 9:389-393.

Baumann, P., L. Baumann, S. S. Bang, and M. J. Woolkalis. 1980. Reevaluation of the taxonomy of *Vibrio, Beneckea* and *Photobacterium*: abolition of the genus *Beneckea. Current Microbiology* 4:127-132.

Bergeron, H. J., Jr. 1987. Method and apparatus for harvesting crawfish and like living things, U.S. Patent 4,663,879, May 12, 1987. U.S. Patent Office, Washington, DC, USA.

Berrill, M., and M. Arsenault. 1984. The breeding behaviour of a northern temperate orconectid crayfish, *Orconectes rusticus. Animal Behavior* 32:333-339.

Berrill, M., and B. Chenoweth. 1981. The burrowing ability of non-burrowing crayfish. *American Midland Naturalist* 108:173-181.

Bills, T. D., and L. L. Marking. 1988. Control of nuisance populations of crayfish with traps and toxicants. *Progressive Fish-Culturist* 50:103-106.

Bishop, J. F. 1968. An ecological study of the branchiobdellid commensals (Annelida-Branchiobdellidae) of some mid-western Ontario crayfish. *Canadian Journal of Zoology* 46:835-843.

Black, J. B. 1966. Cyclic male reproductive activities in the dwarf crawfishes, *Cambarellus shufeldtii* (Faxon) and *Cambarellus puer* Hobbs. *Transactions American Microscopic Society* 85:214-232.

Black, J. B., and J. V. Huner. 1975. Producing your own crayfish stock. *Carolina Tips* 42(4):1-4.

Black, J. B. and J. V. Huner. 1980. Genetics of the red swamp crawfish, *Procambarus clarkii* (Girard): state-of-the-art. *Proceedings of the World Mariculture Society* 11:535-54.

Bodker, J. E., Jr. 1984. Method and apparatus for raising softshell crawfish. U. S. Patent 4,475,480, October 9, 1984.

Borst, D. W. and B. Tsukimura. 1990. Methyl farnesoate levels in hemolymph of crayfish and other crustaceans. *Abstracts, 8th Symposium of Astacology*, Louisiana State University Agricultural Center, Baton Rouge, Louisiana, USA, p. 28.

Bouchard, R. W. 1978. Taxonomy, distribution and general ecology of the genera of North American crayfishes. *The Fisheries Bulletin* 3:11-19.
Bovbjerg, R. 1956. A laboratory culture method for crayfish. *Ecology* 37:613-614.
Broussard, L. J. 1984. Louisiana crawfish farming: pond construction. *American Society of Agricultural Engineers* Paper No. 84-5026, 14 pp.
Brown, P. B., M. L. Hooe, and W. G. Blythe. 1990. Preliminary evaluation of production systems and forages for culture of *Orconectes virilis*, the northern or fantail crayfish. *Journal of the World Aquaculture Society* 21:53-58.
Brown, P. B., M. L. Hooe, and D. H. Buck. 1989. Preliminary evaluation of baits and traps for harvesting orconectid crayfish from earthen ponds. *Journal of the World Aquaculture Society* 20:208-213.
Brown, P. B., C. D. Williams, E. H. Robinson, D. M. Akiyama, and A. L. Lawrence. 1986. Evaluation of methods for determining in vivo digestion coefficients for adult red swamp crayfish *Procambarus clarkii*. *Journal of the World Aquaculture Society* 17:19-24.
Brunson, M. 1989a. Forage and feeding systems commercial crawfish culture. *Journal of Shellfish Research* 8:277-280.
Brunson, M. W. 1989b. *Double cropping crawfish with sorghum in Louisiana.* Louisiana Agricultural Experiment Station, Louisiana State University Agricultural Center, Baton Rouge, Louisiana, USA. Bulletin No. 808.
Brunson, M. W. 1989c. *Evaluation of rice varieties for double cropping crawfish and rice in southwest Louisiana.* Louisiana Agricultural Experiment Station, Louisiana State University Agricultural Center, Baton Rouge, Louisiana, USA. Bulletin No. 812.
Budd, T. W., J. C. Lewis, and M. L. Tracey. 1978. The filter-feeding apparatus in crayfish. *Canadian Journal of Zoology* 56:695-707.
Burns, C., and J. W. Avault, Jr. 1985. Artificial baits for trapping crawfish (*Procambarus* spp.): formulation and assessment. *Journal of the World Mariculture Society* 16:768-374.
Burns, C. M., and J. W. Avault, Jr. 1990. Effectiveness of crawfish baits based on water temperature and bait composition. *Abstracts, 8th International Symposium of Astacology,* Louisiana State University Agricultural Center, Baton Rouge, Louisiana, USA, p. 29.
Busack, C. 1988. Electrophoretic variation in the red swamp (*Procambarus clarkii*) and white river crayfish (*P. sacutus*) (Decapoda: Cambaridae). *Aquaculture* 69:211-226.
Busack, C. A. 1989. Biochemical systematics of crayfishes of the genus *Procambarus*, subgenus *Scapulicambarus* (Decapoda: Cambaridae). *Journal of the North American Benthological Society* 8:180-186.
Caffey, R. H. 1988. An economic analysis of alternative softshell crawfish production facilities. Paper prepared for presentation at the Undergraduate Papers Session, 1988 American Agricultural Economics Association Annual Meeting, Knoxville, Tennessee, USA, July 31-August 3, 1988.
Cain, C. D., Jr., R. A. Bean, and T. French. 1986. Process and apparatus for

harvesting soft shell crayfish. U. S. Patent 4,563,830. U.S. Government Patent Office, Washington, DC, USA. January 14, 1986.

Calala, L. 1976. *Crawfish, keeping crawfish for bait, making softshell crawfish, and their care.* Calala's Water Haven, New London, Ohio, USA.

Cange, S. W., D. Pavel, C. Burns, R. P. Romaire, and J. W. Avault, Jr. 1986a. Evaluation of eighteen artificial crayfish baits. *Freshwater Crayfish* 6:270-273.

Cange, S. W., D. Pavel, and J. W. Avault, Jr. 1986b. Pilot study on prawn/catfish polyculture with rice/crayfish rotation. *Freshwater Crayfish* 6:274-281.

Cange, S. W., C. Burns, J. W. Avault, Jr., and R. P. Romaire (1985-6). Testing artificial baits for trapping crawfish. *Louisiana Agriculture* 29:3-5.

Cange, S. W., M. Miltner, and J. W. Avault, Jr. 1981. Range pellets as supplemental crayfish feed. *Progressive Fish-Culturist* 44:23-24.

Chen, H. M., and S. P. Meyers. 1983. Ensilage treatment of crawfish waste for food improvement of astaxanthin pigment extraction. *Journal of Food Science* 48:1516-1520 & 1555.

Chen, S., J. V. Huner, and R. F. Malone. 1992. Molting and mortality of red swamp and white river crawfish subjected to eyestalk oblation: A preliminary study for commercial soft-shell crawfish production. *Journal of the World Aquaculture Society.* Accepted.

Chien, Y. H., and J. W. Avault, Jr. 1980. Production of crayfish in rice field. *Progressive Fish-Culturist* 42(2):67-71.

Chien, Y. H., and J. W. Avault, Jr. 1983. Effects of flooding dates and disposals of rice straw on crayfish, *Procambarus clarkii* (Girard), culture in rice fields. *Aquaculture* 31:339-350.

Chien, Y. H., and J. W. Avault, Jr. 1979. Double cropping rice, *Oryza sativa* and the red swamp crayfish, *Procambarus clarkii. Freshwater Crayfish* 4:262-272.

Clark, D. E., J. W. Avault, Jr., and S. P. Meyers. 1975. Effects of feeding, fertilization, and vegetation on production of red swamp crayfish, *Procambarus clarkii. Freshwater Crayfish* 2:125-138.

Coker, W. C. 1923. *The Saprolegniaceae with notes on other water molds.* University of North Carolina Press.

Cole, M. T. and M. B. Kilgen. 1987. *Characterization of the quality and shelf-life of whole frozen crawfish.* Final Report for the Gulf and South Atlantic Fisheries Development Foundation, Inc., Tampa, Florida, USA.

Comeaux, M. L. 1975. Historical development of the crayfish industry in the United States. *Freshwater Crayfish* 2:609-620.

Couch, J. A. 1983. Diseases caused by protozoa. Pages 79-111 in A. J. Provenzano (ed.), *The Biology of Crusteacea.* Vol. 6 Pathobiology. Academic Press, New York, USA.

Craft, B. 1980. Some basic considerations in crawfish pond construction. Pages 9-24 in D. Gooch and J. V. Huner (eds.), *First National Crawfish Culture Workshop*, University of Southwestern Louisiana, Lafayette, Louisiana, USA.

Craig, R. J., and W. R. Wolters. 1988. Sources of variation in body size traits, dressout percentage and their correlations for the crayfish, *Procambarus clarkii. Aquaculture* 72:49-58.

Crocker, D. W., and D. W. Barr. 1968. *Handbook of crayfishes of Ontario.* University of Toronto Press, Toronto, Ontario, Canada.

Culley, D. D., and L. F. Duobinis-Gray. 1987a. 24-hour molting pattern of the red swamp crawfish. *Aquaculture* 64:343-346.

Culley, D. D., and L. F. Duobinis-Gray. 1987b. Molting, mortality, and the effect of density in a soft-shell crawfish culture system. *Journal of the World Aquaculture Society* 18:242-246.

Culley, D. D., and L. Duobinis-Gray. 1990. *Culture of the Louisiana soft crawfish. A production manual.* Louisiana Sea Grant Program, Center for Wetland Resources, Louisiana State University, Baton Rouge, Louisiana, USA.

Culley, D. D., Jr., M. Z. Said, and P. T. Culley. 1985. Procedures affecting the production and processing of soft-shelled crawfish. *Journal of the World Mariculture Society* 16:183-192.

Culley, D. D., L. F. Duobinis-Gray, T. B. Lawson, G. R. Baskin, and E. Rejmankova. 1986. Extending the crawfish season. *Louisiana Agriculture* 30:3 & 24.

D'Abramo, L. R., and D. J. Niquette. 1991. Seine harvesting and feeding of formulated feeds as new management practices for pond culture of red swamp crawfish (*Procambarus clarkii* [Girard, 1852]) and white river crawfish, (*P. acutus acutus* [Girard, 1852]) cultured in earthen ponds. *Journal of Shellfish Research* 10:169-178.

D'Abramo, L. R., and E. H. Robinson. 1989. Nutrition of crayfish. *Critical Reviews in Aquatic Sciences* 1:711-728.

Davila, M., and R. Wilkinson. 1990. Growth rates of *Orconectes longedigitus. Program and Abstracts of the 8th International Symposium of Astacology,* Louisiana State University Agricultural Center, Baton Rouge, Louisiana, USA, p. 30.

Davis, B. R., G. R. Fanning, J. M. Madden, A. G. Steigerwalt, H. B. Bradford, H. L. Smith, and D. J. Brenner. 1981. Characterization of biochemically atypical *Vibrio cholerae* strains and designation of a new pathogenic species, *Vibrio mimicus. Journal of Clinical Microbiology* 14(6):631-639.

Davis, D. A., and E. H. Robinson. 1986. Estimation of the dietary lipid requirement level of the white crayfish *Procambarus acutus acutus. Journal of the World Aquaculture Society* 17:37-43.

Day, C. H., and J. W. Avault, Jr. 1986. Crayfish *Procambarus clarkii* production in ponds receiving varying amounts of soybean stubble or rice straw as forage. *Freshwater Crayfish* 6:247-265.

de la Bretonne, L. W., Jr., J. W. Avault, Jr., and R. O. Smitherman. 1969. Effects of soil and water hardness on survival and growth of the red swamp crawfish, *Procambarus clarkii,* in plastic pools. *Proceedings 23rd annual Conference Southeastern Association of Game and Fish Commissioners* 23:626-633.

de la Bretonne, L. W., Jr., and R. P. Romaire. 1989a. Growing crawfish during off-season months. *Crawfish Tales* 8(3):14-16.

FRESHWATER CRAYFISH AQUACULTURE

la Bretonne, L. W., Jr., and R. P. Romaire. 1989b. Commercial crawfish cultivation practices: A review. *Journal of Shellfish Research* 8:267-276.

Delibes, M., and I. Adrian. 1987. Effects of crayfish introduction on otter *Lutra lutra* food in the Donana National Park, SW Spain. *Biological Conservation* 42:153-159.

Dellenbarger, L. E., and E. J. Luzar. 1988. The economics associated with crawfish production from Louisiana's Atchafalaya Basin. *Journal of the World Aquaculture Society* 19:41-46.

Dellenbarger, L. E., K. J. Roberts, S. S. Kelly, and P. W. Pawlyk. 1986. An analysis of the Louisiana crawfish processing industry and potential market outlets. Department of Agricultural Economics and Agribusiness, Louisiana State University, Baton Rouge, Louisiana, USA. *D.A.E. Research Report* No. 654.

Dellenbarger, L. E., L. R. Vandeveer, and T. M. Clarke. 1987. Estimated investment requirements, production costs, and breakeven prices for crawfish in Louisiana, 1987. *Department of Agricultural Economics and Agribusiness Research Report* No. 670, Louisiana State University Agricultural Center, Baton Rouge, Louisiana, USA.

Dellenbarger, L. E., A. R. Schupp, and H. O. Zapata. 1990. Consumer and grocery store experience with crawfish in four selected U.S. cities. Department of Agricultural Economics and Agribusiness, Louisiana State University, Baton Rouge, Louisiana, USA. *A.E.A. Information Series* No. 81.

Dellenbarger, L. E., A. R. Schupp, and H. O. Zapata. 1990-91. Targeting metropolitan markets out of state for crawfish. *Louisiana Agriculture* 34(2):12-13.

Desselle, L. J. 1980. The value of crawfish waste as a lime source and soil amendment. Pages 77-83 in D. Gooch and J. V. Huner (eds.), *Proceedings First National Crawfish Culture Workshop*, University of Southwestern Louisiana, Lafayette, Louisiana, USA.

Dye, L. and P. Jones. 1975. The influence of density and invertebrate predation on the survival of young-of-the-year *Orconectes virilis*. *Freshwater Crayfish* 2:529-538.

Ekanem, S. B., J. W. Avault, Jr., J. B. Graves, and H. Morris. 1981. Acute toxicity of propanil, ordram, and furadan to crawfish (*Procambarus clarkii*) when chemicals were combined and used alone. *Journal of the World Mariculture Society* 12(2):373-383.

Ekanem, S. B., J. W. Avault, Jr., J. B. Graves, and H. Morris. 1983. Effects of rice pesticides on *Procambarus clarkii* in a rice/crawfish pond model. *Freshwater Crayfish* 5:315-323.

Eversole, A. G. 1990. Diversification of crawfish management schedule. *Journal of the World Aquaculture Society* 21:59-63.

Eversole, A. G., and R. S. Pomeroy. 1989. Crawfish culture in South Carolina: an emerging aquaculture industry. *Journal of Shellfish Research* 8:309-313.

Feminella, J. W., and V. H. Resh. 1989. Submersed macrophytes and grazing crayfish: an experimental study of herbivory in a California freshwater marsh. *Holarctic Ecology* 12:1-8.

Finerty, M. W., J. D. Madden, S. E. Feagley, and R. M. Grodner. 1990. Effect of environs and seasonality on metal residues in tissues of wild and pond-raised crayfish in southern Louisiana. *Archives of Environmental Contamination Toxicology* 19:94.

Fisher, W. S. 1977. Epibiotic microbial infestations of cultured crustaceans. *Proceedings of the World Mariculture Society* 8:673-684.

Forney, J. L. 1958 (revised 1968). Raising bait fish and crayfish in New York ponds. *Cornell Extension Bulletin 986*, Cornell University, Ithaca, New York, USA.

Gaudé, A. P., III. 1982. The impact of naiad dragonflies (*Anax junius* Drury) on juvenile crawfish (*Procambarus* spp.) in commercial crawfish culture. *Completion Report*, Special Research Project, Department of Biology, University of Southwestern Louisiana, Lafayette, Louisiana, USA. 15 pp. (Mimeo).

Gaudé, A. P., III. 1988. Thermal effects on pesticide toxicity for Louisiana red swamp crawfish (*Procambarus clarkii*). *Freshwater Crayfish* 7:171-177.

Gaudé, A. P., III, and G. M. Gaudé. 1989. Grading crawfish. *Crawfish Tales* 8(1):20-23.

Goddard, J. S. 1988. Food and feeding. Pages 145-166 in D. M. Holdich and R. S. Lowery (eds.), *Freshwater Crayfish: Biology, Management and Exploitation*. Croom Helm, London, England and Sydney, Australia.

Goyert, J. C., and J. W. Avault, Jr. 1977. Agricultural by-products as supplemental feed for crayfish, *Procambarus clarkii*. *Transactions of the American Fisheries Society* 106:629-633.

Goyert, J. C., and J. W. Avault, Jr. 1978. Effects of stocking density and substrate on growth and survival of crawfish (*Procambarus clarkii*) grown in a recirculating system. *Proceedings of the World Mariculture Society* 9:731-735.

Goyert, J. C., and J. W. Avault, Jr. 1979. Effects of container size of growth of crawfish (*Procambarus clarkii*) in a recirculating system. *Freshwater Crayfish* 4:277-286.

Grabda, E. 1934. Récherches sur un parasite de l'écrévisse (*Potamobius fluviatilis* L.), connu sous le nom de *Psorospermium haeckeli* Hlgd. - *Mem. de l'acad. Pol. des sciences et des lettres. S.B., Math. u. Naturw.* 6:123-142. Cracovie.

Green, L. M., J. S. Tuten, and J. W. Avault, Jr. 1979. Polyculture of red swamp crawfish (*Procambarus clarkii*) and several North American fish species. *Freshwater Crayfish* 4:287-298.

Hall, T. J. 1979. Ectocommensals of the freshwater shrimp, *Macrobrachium rosenbergii*, in culture facilities at Homestead, Florida. Pages 214-219 in *Proc. Second Biennial Crustacean Health Workshop*, Texas A & M. Sea Grant College Program, College Station, Texas, USA. Publication No. TAMU-SG-79-114.

Hamajima, F., F. Fujino, and M. Koga. 1976. Studies on the host-parasite relationship of *Paragonimus westerman*: (Kerbert, 1878). IV. Predatory habits of some freshwater crabs and crayfish on the snail *Semisalcaspira libertina* (Gould, 1859). *Annotationes Zoologicae Japoneyes* 49:224-278.

Hazlett, B. A., and D. Rittschof. 1985. Variation in rate of growth in the crayfish *Orconectes virilis*. *Journal of Crustacean Biology* 5:341-346.

Hentonnen, P., O. V. Lindqvist, and J. V. Huner. 1992. Incidence of *Psorospermium* sp. in several cultivated populations of crayfishes, *Procambarus* spp. in Southern Louisiana. *Journal of the World Aquaculture Society* 23:31-37.

Hobbs, H. H., Jr. 1972. Biota of freshwater ecosystems. *Identification manual 9: Crayfishes (Astacidae) of North and Middle America*. Water Pollution Control Series. U. S. Environmental Protection Agency, Washington, DC, USA.

Hobbs, H. H., Jr. 1974. A checklist of the North and Middle American crayfishes (Decapoda: Astacidae and Cambaridae). *Smithsonian Contributions to Zoology* 166:1-161.

Hobbs, H. H., Jr. 1975. Adaptations and convergence in North American crayfish. *Freshwater Crayfish* 2:541-553.

Hobbs, H. H., Jr. 1981. The crayfishes of Georgia. *Smithsonian Contributions to Zoology* 318:1-549.

Hobbs, H. H., Jr. 1988. Crayfish distribution, adaptive radiation and evolution. Pages 52-82 in D. M. Holdich and R. S. Lowery (eds.), *Freshwater Crayfish: Biology, Management and Exploitation*. Croom Helm, London, England and Sydney, Australia.

Hobbs, H. H., Jr. 1989. An illustrated checklist of the American crayfishes (Decapoda: Astacidae, Cambaridae, and Parastacidae). *Smithsonian Contributions to Zoology* 480:1-236.

Hobbs, H. H., Jr., and H. H. Hobbs, III. 1990. A new crayfish (Decapoda: Cambaridae) from southeastern Texas. *Proceedings Biological Society of Washington* 103:608-613.

Hobbs, H. H., III, and J. P. Jass. 1988. *The crayfishes and shrimp of Wisconsin*. Milwaukee Public Museum, Milwaukee, Wisconsin, USA.

Hobbs, H. H., III, J. Jass, and J. V. Huner. 1989. A review of global crayfish introductions with particular emphasis on two North American species (Decapoda, Cambaridae). *Crustaceana* 56:299-316.

Holt, Perry C. 1973. A free-living branchiobdellid (Annelida: Clitellata). *Transactions of the American Microscopical Society* 92(1):152-153.

Holt, P. S. 1975. The branchiobdellid (Annelida: Clitellata) associates of astocoidean crawfishes. *Freshwater Crayfish* 2:332-345.

Hubbard, D. M., E. H. Robinson, P. B. Brown, and W. H. Daniels. 1986. Optimum ratio of dietary protein to energy for red crayfish (*Procambarus clarkii*). *Progressive Fish-Culturist* 48:233-237.

Huner, J. V. 1975. Observations on the life histories of recreationally important crawfishes in temporary habitats. *Proceedings of the Louisiana Academy of Sciences* 38:20-24.

Huner, J. V. 1976. Raising crawfish for fish bait and food: a new polyculture crop with fish. *The Fisheries Bulletin* 1:7-8.

Huner, J. V. 1978a. Exploitation of freshwater crayfishes in North America. *The Fisheries Bulletin* 3:2-5.

Huner, J. V. 1978b. Crawfish population dynamics as they affect production in several small, open crawfish ponds in Louisiana. *Proceedings of the World Mariculture Society* 9:619-640.

Huner, J. V. 1984. Crawfish in Ohio. *Crawfish Tales* 3(1):20-22.

Huner, J. V. 1986a. Cropping fish and crawfish in the same pond. *Crawfish Tales* 5(1):21-23.

Huner, J. V. 1986b. Cottonseed cake: an effective crawfish bait and turtle deterrent. *Crawfish Tales* 5(1):34-35.

Huner, J. V. 1987. Tolerance of the crawfishes *Procambarus acutus acutus* and *Procambarus clarkii* (Decapoda, Cambaridae) to acute hypoxia and elevated thermal stress. *Journal of the World Aquaculture Society* 18:113-114.

Huner, J. V. 1988a. Comparison of the morphology and meat yield of red swamp crawfish and white river crawfish. *Crawfish Tales* 7:2.

Huner, J. V. 1988b. Crawfish for fish bait. The Louisiana perspective. *Crawfish Tales* 7(3):12-14.

Huner, J. V. 1988c. Soft-shell crawfish industry. Pages 28-42 in L. Evans and D. O'Sullivan (eds.), *Proceedings, First Australian Crustacean Aquaculture Conference*, Curtin University of Technology, Perth, Australia.

Huner, J. V. 1989a. Crawfish–the U.S. picture. *Seafood International* 4(9):53-56.

Huner, J. V. 1989b. Survival of red swamp and white river crawfishes under simulated burrow conditions. *Crawfish Tales* 8(1):29.

Huner, J. V. 1989c. Overview of international and domestic freshwater crawfish production. *Journal of Shellfish Research* 8:259-266.

Huner, J. V. 1989d. Automated crawfish harvesting machines: a status report. *Crawfish Tales* 8(3):20.

Huner, J. V. 1990a. Use of seines and harvesting machines in soft-shell crawfish industry. *Crawfish Tales* 9(1):16.

Huner, J. V. 1990b. Wading bird in crawfish, a new problem? *Crawfish Tales* 9(1):20.

Huner, J. V. 1990c. Fuzzy crawfish. *Crawfish Tales* 9(2):24-25.

Huner, J. V. 1990d. Biology, fisheries, and cultivation of freshwater crawfishes in the U.S. *Reviews in Aquatic Sciences* 2:229-254.

Huner, J. V. 1992. Experiences with the supplemental stocking of red swamp, *Procambarus clarkii*, and white river, *Procambarus zonangulus*, crayfishes for growout purposes. *Finnish Fisheries Research*. In Press.

Huner, J. V. and G. R. Abraham. 1983. Observations on wading birds. *Crawfish Tales* 2:16-19.

Huner, J. V., and J. W. Avault, Jr. 1976a. The molt cycle of subadult red swamp crawfish, *Procambarus clarkii* (Girard). *Proceedings of the World Mariculture Society* 7:267-273.

Huner, J. V., and J. W. Avault, Jr. 1976b. Sequential pond flooding: a prospective management technique for extended production of bait size crawfish. *Transactions of the American Fisheries Society* 105:637-642.

Huner, J. V., and J. W. Avault, Jr. 1977. Investigation of methods to shorten the intermolt period in a crawfish. *Proceedings of the World Mariculture Society* 8:883-893.

Huner, J. V., and J. E. Barr. 1991 (3rd revision). *Red swamp crawfish: biology and exploitation.* Sea Grant No. LSU-T-80-001, LSU Center for Wetland Resources, Baton Rouge, Louisiana, USA.

Huner, J. V., and A. P. Gaudé III. 1989. Hatcheries for red swamp crawfish. *Crawfish Tales* 8(1):23-25.

Huner, J. V., and M. J. LeBlanc. 1991. Raising catfish in Louisiana crawfish ponds has potential. *The Catfish Journal* 5(6):25&27.

Huner, J. V., and S. P. Meyers. 1979. Dietary protein requirements of the red swamp crawfish, *Procambarus clarkii*, grown in a closed system. *Proceedings of the World Mariculture Society* 10:751-760.

Huner, J. V., and S. Naqvi. 1984. Invertebrate faunas and crawfish food habits in Louisiana crawfish ponds. *Proceedings Annual Conference of Southeastern Fish and Wildlife Agencies* 38:395-406.

Huner, J. V., and J. Paret. 1990. Seasonal changes in species composition of various baits in a south Louisiana commercial crawfish (*Procambarus* spp.) pond. *Program and Abstracts of the 8th International Symposium of Astacology*, Louisiana State University Agricultural Center, Baton Rouge, Louisiana, USA, p. 39.

Huner, J. V., and V. A. Pfister. 1990. Feasibility of stocking juvenile crawfish in small ponds. *Crawfish Tales* 9(1):13-14.

Huner, J. V., and R. P. Romaire. 1979. Size at maturity as a means of comparing populations of *Procambarus clarkii* (Girard) (Crustacea: Decapoda) from different habitats. *Freshwater Crayfish* 4:53-64.

Huner, J. V., J. Kowalczuk, and J. W. Avault, Jr. 1976. Calcium and magnesium levels in the intermolt (C_4) carapaces of three species of freshwater crawfish (Cambaridae: Decapoda). *Comparative Biochemistry and Physiology* 55A: 183-186.

Huner, J. V., J. G. Kowalczuk, and J. W. Avault, Jr. 1978. Postmolt calcification in subadult red swamp crayfish, *Procambarus clarkii* (Girard) (Decapoda, Cambaridae). *Crustaceana* 34:275-280.

Huner, J. V., O. V. Lindqvist, and H. Könönen. 1988. Comparison of morphology and edible tissues of two important commercial crayfishes, the noble crayfish, *Astacus astacus* Linn., and the red swamp crayfish, *Procambarus clarkii* (Girard) (Decapoda, Astacidae and Cambaridae). *Aquaculture* 68:45-57.

Huner, J. V., M. Miltner, and J. W. Avault, Jr. 1983a. Crawfish, *Procambarus* spp., production from summer flooded experimental ponds used to culture prawns, *Macrobrachium rosenbergii*, and/or channel catfish, *Ictalurus punctatus*, in south Louisiana. *Freshwater Crayfish* 5:379-390.

Huner, J. V., R. P. Romaire, and L. W. de la Bretonne, Jr. 1989. Crawfish traps and baits. *Crawfish Tales* 8(2):12-14.

Huner, J. V., R. Malone, and M. Fingerman. 1990. Practical application of eyestalk ablation for producing soft-shell crawfish. *Abstracts, 8th International Symposium of Astacology*, Louisiana State University Agricultural Center, Baton Rouge, Louisiana, USA, p. 38.

Huner, J. V., R. P. Romaire, V. Pfister, and T. Baum. 1991. Effectiveness of commercially formulated and fish baits in trapping cambarid crawfish. *Journal of the World Aquaculture Society* 21:288-294.

Huner, J. V., M. Miltner, J. W. Avault, Jr., and R. A. Bean. 1983b. Interaction of freshwater prawns, channel catfish fingerlings, and crayfish in earthen ponds. *Progressive Fish-Culturists* 45:36-40.

Hymel, T. M. 1987. Water quality in crawfish ponds. Pages 74-76 in R. Reigh (ed.), *Proceedings 1st Louisiana Aquaculture Conference*, Louisiana Cooperative Extension Service, Louisiana State University, Baton Rouge, Louisiana, USA.

Hymel, T. M. 1988. Paddlewheel aerator update. *Crawfish Tales* 7(4):22-24.

Jarboe, J. H. 1988. The toxicity of pesticides to crawfish. *Crawfish Tales* 7(4): 25-29.

Jaspers, E., and J. W. Avault, Jr. 1969. Environmental conditions in burrows and ponds of the red swamp crawfish, *Procambarus clarkii* (Girard), near Baton Rouge, Louisiana. *Proceedings 23rd Annual Conference of the Southeastern Association of Game and Fisheries Commissioners* 23:634-647.

Johnson, S. K. 1977. *Crawfish and freshwater shrimp diseases.* Texas A & M University Sea Grant College Program. Texas Agricultural Extension Service, College Station, Texas, USA. Pbl. No. TAMU-SG-77-605.

Johnson, W. B., Jr., L. L. Glasgow, and J. W. Avault, Jr. 1983. A comparison of delta duck potato (*Sagittaria graminea platyphylla*) and rice (*Oryza sativa*) as cultured red swamp crayfish (*Procambarus clarkii*) forage. *Freshwater Crayfish* 5:351:361.

Kartamulia, K., and D. B. Rouse. 1991. Survival and growth of marron, *Cherax tenuimanus*, in outdoor tanks in the southeastern USA. *Journal of the World Aquaculture Society* 23(2): In Press.

Klaassen, H. E. 1986. Potential crayfish culture in Kansas. *Kansas Commercial Fish Growers Association Newsletter*, October-November 1986:1-4.

Kossakowski, J. 1966. *Crayfish (Raki).* Panstowe Wydownectwo Polnicze i Lesne. Warsaw, Poland. Available from US Department of Commerce, Washington, DC, USA. Translation 77-55114 (Translated from Polish, 1971).

Lalla, H., and T. B. Lawson. 1987. Depuration of crawfish with a water spray. *American Society of Agricultural Engineers Paper No. 87-5034.*

Langlois, T. H. 1935. Notes on the habits of the crayfish, *Cambarus rusticus* Girard, in fish ponds in Ohio. *Transactions of the American Fisheries Society* 65:189-192.

Laufer, H., M. Landau, D. Borst, and E. Homola. 1986. The synthesis and regulation of methyl farnesoate, a new juvenile hormone for crustacean reproduction. *Advances in Invertebrate Reproduction* 4:135-143.

Lawson, T. B. 1990. Aeration in crawfish ponds. Pages 55-60 in R. C. Reigh (ed.), *Proceedings Louisiana Aquaculture Conference*, Louisiana State University Agricultural Center, Baton Rouge, Louisiana, USA.

Lawson, T. B., G. R. Baskin, and J. D. Bankston. 1984. Non-mechanical aerating device for crawfish ponds. *American Society of Agricultural Engineers Paper No. 84-5027.* 13 pp.

Lawson, T. B., and G. R. Baskin. 1985. Crawfish holding and purging systems. *American Society of Agricultural Engineers Paper No. 85-5008.* 17 pp.

Lawson, T. B., H. Lalla, and R. P. Romaire. 1990. Purging crawfish in a water spray system. *Journal of Shellfish Research* 9:383-387.

Lee, B. J., R. F. Sis, D. H. Lewis, and J. E. Marks. 1985. Histology of select organs of the crawfish *Procambarus clarkii* maintained at various temperatures and levels of calcium and ammonia. *Journal of the World Mariculture Society* 16:193-204.

Leonhard, S. L. 1981. *Orconectes virilis.* In S. G. Lawrence (ed.), Manual for the Culture of Selected Freshwater Invertebrates. *Canadian Special Publications of Fish and Aquatic Sciences* 54:95-108.

Leung, T., S. M. Naqvi, and N. S. Naqvi. 1980. Paraquat toxicity of Louisiana crayfish. *Bulletin of Environmental Contamination and Toxicology* 25:465.

Lindqvist, O. V., and H. Mikkola. 1979. On the etiology of the muscle wasting disease in *Procambarus clarkii* in Kenya. *Freshwater Crayfish* 4:363-372.

Little, E. E. 1975. Chemical communication in maternal behavior of crayfish. *Nature* 255:400-401.

Little, E. E. 1976. Ontogeny of maternal behavior and brood pheromone in crayfish. *Journal of Comparative Physiology* 112:133-142.

Ljungberg, O., and L. Monné. 1968. On the eggs of anenigmatic nematode parasite encapsulated in the connective tissue of the European crayfish *Astacus* in Sweden. *Bulletin of the Office of International Epizootics* 69:1231-1235.

Lodge, D. M., A. L. Beckel, and J. J. Magnuson. 1985. Lake-bottom tyrant. *Natural History Magazine* 85/8:32-37.

Lorman, J. G., and J. J. Magnuson. 1978. The role of crayfish in aquatic ecosystems. *The Fisheries Bulletin* 3:8-10.

Lowery, R. S., and A. J. Mendes. 1977. *Procambarus clarkii* in Lake Naivasha, Kenya, and its effects on established and potential fisheries. *Aquaculture* 11: 111-121.

Loyacano, H. 1967. Some effects of salinity on two populations of red swamp crawfish, *Procambarus clarkii. Proceedings Annual Conference Southeastern Association of Game and Fish Commissioners* 21:423-434.

Lutz, C. G., and W. R. Wolters. 1986. The effect of five stocking densities on growth and yield of red swamp crawfish *Procambarus clarkii. Journal of the World Aquaculture Society* 17:33-36.

Lutz, C. G., and W. R. Wolters. 1987. Growth and yield of red swamp crawfish and white river crawfish stocked separately and in combination. *Crawfish Tales* 6(2):20-21.

Lutz, C. G., and W. R. Wolters. 1989. Estimation of heritabilities for growth, body size, and processing traits in red swamp crawfish, *Procambarus clarkii* (Girard). *Aquaculture* 78:21-33.

Madden, J. D., M. W. Finerty, and R. M. Grodner. 1989. Survey of persistent pesticide residues in the edible tissues of wild pond-raised Louisiana crayfish and their habitat. *Bulletin of Environmental Contamination and Toxicology* 43:779.

Malone, R. F., and D. G. Burden. 1988. *Design of recirculating soft crawfish shedding systems.* Louisiana Sea Grant College Program, Center for Wetland Resources, Louisiana State University, Baton Rouge, Louisiana, USA.

Malone, R. F., and D. D. Culley. 1988. Method and apparatus for farming softshell aquatic crustaceans. U.S. Patent No. 4,726,321. U.S. Patent Office, Washington, DC, February 23, 1988.

Malone, R. F., J. E. Robin, and D. E. Coffin. 1991. Low density biomedia filtration of a commercial recirculating soft crawfish production facility. *Program and Abstracts, 22nd Annual Conference & Exposition, World Aquaculture Society,* San Juan, Puerto Rico, p. 41.

Marshall, G. A. 1988. Processing and freezing methods influencing the consistency and quality of fresh and frozen peeled crawfish (*Procambarus* sp.) meat. Ph.D. dissertation. Louisiana State University, Baton Rouge, Louisiana, USA.

Marshall, G. A., and M. W. Moody. 1986. Department of Food Science, Louisiana State University, Baton Rouge, Louisiana, USA. Unpublished data.

Marshall, G. A., M. W. Moody, C. R. Hackney, and J. S. Godber. 1987. Effect of blanch time on the development of mushiness in the ice-stored crawfish meat packed with adhering hepatopancreas. *Journal of Food Science* 52:6.

Marshall, G. A., M. W. Moody, and C. R. Hackney. 1988. Differences in color, texture, and flavor of processed meat from red swamp crawfish (*Procambarus clarkii*) and white river crawfish (*P. acutus acutus*). *Journal of Food Science* 53(1):280.

Martin, R. P., and R. B. Hamilton. 1985. Wading bird predation in crawfish ponds. *Louisiana Agriculture* 28(4):3-5.

McCartney, B., and J. W. Garrett. 1989. *Temperate crayfish culture.* McCartney and Garrett, El Dorado Springs, Missouri, USA.

McClain, W. R., and P. K. Bollich. 1992. Relaying: A means to increase marketing size of Louisiana crayfish. *Abstracts of Papers. Ninth International Symposium of Astacology.* Reading, England, p. 12.

McMahon, B. R. 1986. The adaptable crayfish: mechanisms of physiological adaptation. *Freshwater Crayfish* 6:59-74.

McMahon, B. R. and S. A. Stuart. 1990. Physiological effects of long-term air exposure in *Procambarus clarkii. Abstracts, 8th International Symposium of Astacology.* Louisiana State University Agricultural Center, Baton Rouge, Louisiana, USA, p. 48.

McNease, L. and T. Joanen. 1977. Alligator diets in relation to marsh salinity. *Proceedings Annual Meeting Southeastern Association of Game and Fish Commissioners* 31:36-40.

Medley, P. B., L. U. Hatch, R. G. Nelson, and D. B. Rouse. 1991a. Economic feasibility and risk analysis of pond produced Australian red claw crayfish (*Cherax quadricarinatus*) in the southeastern United States. *Program and Abstracts, 22nd Annual Conference & Exposition,* World Aquaculture Society, San Juan, Puerto Rico, p. 44.

Medley, P. B., D. B. Rouse, and Y. J. Brady. 1991b. Biological interactions of the red swamp crayfish (*Procambarus clarkii*) and the Australian red claw (*Cher-*

ax quadricarinatus) in communal culture ponds. *Program and Abstracts, 22nd Annual Conference & Exposition*, World Aquaculture Society, San Juan, Puerto Rico, p. 44.

Meyers, S. P. 1985. Recovery of meat from crawfish claws. *Crawfish Tales* 4(2):28-29.

Meyers, S. P. 1987. Crawfish-total product utilization. *Infofish Marketing Digest* 3/87, May/June:31.

Meyers, S. P., and D. Bligh. 1981. Characterization of astaxanthin pigments from heat-processed crawfish waste. *Journal Agricultural and Food Chemistry* 29:505-508.

Miltner, M. R., and J. W. Avault, Jr. 1983. An appropriate food delivery system for low levee pond culture of *Procambarus clarkii*, the red swamp crayfish. *Freshwater Crayfish* 5:379-390.

Moody, M. W. 1989. Processing of freshwater crawfish: a review. *Journal of Shellfish Research* 8:293.

Moody, M. W. and G. M. Moertle. 1986. An evaluation of the effectiveness of specific chemical compounds on the prevention of discoloration development during the cooking of frozen crawfish meat. Annual Meeting of the Institute of Food Technologist, Dallas, Texas, USA.

Momot, W. T. 1984. Crayfish production: a reflection of community energetics. *Journal of Crustacean Biology* 4:35-54.

Momot, W. T. 1988. *Orconectes* in North America and elsewhere. In pp. 262-283, D. M. Holdich and R. S. Lowery (eds.), *Freshwater Crayfish: Biology, Management and Exploitation*. Croom Helm, London, England and Sydney, Australia.

Momot, W. T. and R. P. Romaire. 1981. Use of a seine to detect stunted crawfish populations in a pond, a preliminary report. *Journal World Mariculture Society* 12(2):384-390.

Momot, W. T., H. Gowing, and P. D. Jones. 1978. The dynamics of crayfish and their role in the ecosystem. *American Midland Naturalist* 99:10-35.

Morgan, K. L., R. J. Edling, and J. A. Musick. 1982-83. A mechanical crawfish harvester. *Louisiana Agriculture* 26(2):4-5.

Morrissy, N. M. 1979. Experimental pond production of marron, *Cherax tenuimanus* (Smith). *Aquaculture* 16:319-344.

Murofushi, M., Y. Deguchi, and T. H. Yosida. 1984. Karyological study of the red swamp crayfish and the Japanese lobster by air-drying method. *Proceedings Japanese Academy*, Series B. 60:306-309.

Naqvi, S. M., R. Hawkins, and N. H. Naqvi. 1987. Mortality response and LC_{50} for juvenile and adult crayfish, *Procambarus clarkii* exposed to Thiodan®, MSMA® (herbicides) and Cutrine-Plus® (algicide). *Journal of Environmental Pollution* 48:275-283.

Naqvi, S. M., C. T. Flagge, and R. L. Hawkins. 1990. Arsenic uptake and depuration by red crayfish, *Procambarus clarkii*, exposed to various concentrations of Monosodium Methanearsonate (MSMA) herbicide. *Bulletin of Environmental Contamination and Toxicology* 45:94-100.

Naqvi, S. M., and C. T. Flagge. 1990. Chronic effects of arsenic on American red crayfish, *Procambarus clarkii*, exposed to Monosodium Methanearsonate (MSMA) herbicide. *Bulletin of Environmental Contamination and Toxicology* 45:101-106.

Nassar, J. R., P. J. Zwank, D. C. Hayden, and J. V. Huner. 1991. *Construction and management of multiple-use impoundments for waterfowl and crawfish.* U.S. Department of the Interior, Fish & Wildlife Service, National Wetlands Research Center, Slidell, Louisiana, USA, 48 pp.

Niquette, D. J. and L. R. D'Abramo. 1989. Use of a seine to harvest and monitor population dynamics in intensive crawfish culture. *Crawfish Tales* 8(2):15-17.

Niquette, D. J. and L. R. D'Abramo. 1991. Population dynamics of red swamp crawfish, *Procambarus clarkii* (Girard, 1852), and white river crawfish, *P. acutus acutus* (Girard, 1852), cultured in earthen ponds. *Journal of Shellfish Research* 10:179-186.

No, H. K., and S. P. Meyers. 1989a. Crawfish chitosan as a coagulant in recovery of organic compounds from seafood processing streams. *Journal of Agriculture and Food Chemistry* 37:580.

No, H. K., and S. P. Meyers. 1989b. Recovery of amino acids from seafood processing wastewater with a dual chitosan-based ligand-exchange system. *Journal of Food Science* 54(1):60.

No, H. K., S. P. Meyers, and K. S. Lee. 1989. Isolation and characterization of chitin from crawfish shell waste. *Journal of Agriculture and Food Chemistry* 37:575.

Nylund, V., K. Westman, and K. Lounatmaa. 1983. Ultrastructure and taxonomic position of the crayfish parasite *Psorospermium haeckeli* Hilgendorf. *Freshwater Crayfish* 5:307-314.

Palva, T. K., and J. V. Huner. 1989. Analysis of freshwater crayfish mitochondrial DNA: a short report. *World Aquaculture* 19(4):82.

Patrick, R. M., and M. W. Moody. 1989. *Enjoying Louisiana crawfish.* Louisiana Cooperative Extension Service Pub. 2353, Louisiana State University, Baton Rouge, Louisiana, USA, 4 pp.

Payne, J. F. 1978. Aspects of the life histories of selected species of North American crayfishes. *The Fisheries Bulletin* 3:5-8.

Penn, G. H., Jr. 1943. A study of the life history of the Louisiana red crawfish, *Cambarus clarkii* Girard. *Ecology* 24:1-18.

Penn, G. H. 1950. Utilization of crawfishes by coldblooded vertebrates in the eastern USA. *American Midland Naturalist* 44:643-658.

Penn, G. H. 1956. The genus *Procambarus* in Louisiana (Decapoda, Astacidae). *American Midland Naturalist* 56:406-422.

Perry, W. G., Jr., and C. G. LaCaze. 1969. Preliminary experiment on the culture of red swamp crawfish, *Procambarus clarkii*, in brackish water ponds. *Proceedings 23rd Annual Conference Southeastern Association of Game and Fish Commissioners* 23:251-254.

Perry, W. G., Jr., T. Joanen, and L. McNease. 1970. Crawfish-waterfowl, a multiple use concept for impounded marshes. *Proceedings 24th Annual Conference of Southeastern Association of Game and Fish Commissioners* 24:506-519.

Pfister, V. E. and R. P. Romaire. 1983. Catch efficiency and retentive ability of commercial crawfish traps. *Aquacultural Engineering* 2:107-118.

Pomeroy, R. S., D. B. Luke, and J. Whetstone. 1989. *Budgets and cash flow statements for South Carolina crawfish production.* Clemson University Cooperative Extension Service Extension Economics Report EER 106, Clemson University, Clemson, South Carolina, USA.

Price, J. O. and J. F. Payne. 1979. Multiple summer moults in adult *Orconectes neglectus chenodactylus. Freshwater Crayfish* 4:93-104.

Quinn, J. R. 1989. Crawfish in the aquarium. *Tropical Fish Hobbyist* 37(11): 34-41.

Rach, J. J., and T. D. Bills. 1987. Comparison of three baits for trapping crayfish. *North American Journal of Fisheries Management* 7:601-603.

Rach, J. J., and T. D. Bills. 1989. Crayfish control with traps and largemouth bass. *Progressive Fish-Culturist* 51:157-160.

Reigh, R. C., S. L. Braden, and R. J. Craig. 1990. Apparent digestibility coefficients for common feedstuffs in formulated diets for red swamp crayfish, *Procambarus clarkii. Aquaculture* 84:321-334.

Reitz, R. C., J. W. Wilson, D. Culley, and G. W. Winston. 1990. The fatty acids and cholesterol content of soft shell crayfish, *Procambarus clarkii. Program and Abstracts of the 8th International Symposium of Astacology,* Louisiana State University Agricultural Center, Baton Rouge, Louisiana, USA, p. 54.

Riché, J. 1989. Aquaculture harvester for shallow-water use. U.S. Patent No. 4,813,377. U.S. Patent Office, Washington, DC, March 21, 1989.

Rickett, J. D. 1974. Trophic relationships involving crayfish of the genus *Orconectes* in experimental ponds. *Progressive Fish-Culturist* 36:207-211.

Roberts, K. J., and L. Dellenbarger. 1989. Louisiana crawfish product markets and marketing. *Journal of Shellfish Research* 8:303-309.

Robin, J. E., S. Chen, and R. F. Malone. 1991. Operational and management strategies for a commercial-scale automated soft crawfish production facility. *Program and Abstracts, 22nd Annual Conference & Exposition,* World Aquaculture Society, San Juan, Puerto Rico, p. 54.

Rogers, R., and J. V. Huner. 1985. Comparison of burrows and burrowing behavior of five species of cambarid crawfish from the Southern University campus, Baton Rouge, Louisiana. *Proceedings of the Louisiana Academy of Sciences* 48:23-29.

Romaire, R. P. 1988. Trap designs and catchability. *Crawfish Tales* 7(1):35-37.

Romaire, R. P. 1989a. Use of formulated feeds in crawfish culture. *Crawfish Tales* 8(3):28-32.

Romaire, R. P. 1989b. Overview of harvest technology used in freshwater crawfish culture. *Journal of Shellfish Research* 8:281-286.

Romaire, R. P. 1990. *Crawpop-red swamp crawfish aquaculture simulation model.* School of Forestry, Wildlife and Fisheries, Louisiana State University, Baton Rouge, Louisiana, USA.

Romaire, R. P., and C. G. Lutz. 1990. Population dynamics of *Procambarus clarkii* (Girard) and *Procambarus acutus acutus* (Girard) (Decapoda: Cambaridae) in commercial ponds. *Aquaculture* 81:253-274.

Romaire, R. P., and V. H. Osorio. 1986. Evaluation of manufactured baits. *Crawfish Tales* 5(4):24-29.

Romaire, R. P., and V. H. Osorio. 1989. Effectiveness of crawfish baits as influenced by habitat type, trap-set time, and bait quantity. *Progressive Fish-Culturist* 51:232-237.

Romaire, R. P., and V. A. Pfister. 1983a. Evaluating crawfish traps. *Louisiana Agriculture* 26:3 & 24.

Romaire, R. P., and V. A. Pfister. 1983b. Effects of trap density and diel harvesting frequency on catch of crawfish. *North American Journal of Fisheries Management* 3:419-424.

Romaire, R. P., J. Forester, and J. W. Avault, Jr. 1979. Growth and survival of stunted red swamp crawfish (*Procambarus clarkii*) in a feeding-stocking density experiment in pools. *Freshwater Crayfish* 4:331-336.

Rondelle, R. F., S. Chen, and R. F. Malone. 1991. Vertical distribution in an automated recirculating soft-shell crawfish separator. *Program and Abstracts, 22nd Annual Conference & Exposition*, World Aquaculture Society, San Juan, Puerto Rico, p. 55.

Rouse, D. B. and P. B. Medley. 1991. Experimental pond production of the Australian red claw (*Cherax quadricarinatus*) in the southeastern United States. *Program and Abstracts, 22nd Annual Conference & Exposition*, World Aquaculture Society, San Juan, Puerto Rico, pp. 55-56.

Sakazaki, R., and T. Shimada. 1977. Serovars of *Vibrio cholerae. Japanese Journal of Medical Scientific Biology* 30:279-282.

Sandifer, P. A. 1988. Aquaculture in the West, a perspective. *Journal of the World Aquaculture Society* 19:73-84.

Sanguanruang, M. 1988. Bioenergetics of red swamp crawfish (*Procambarus clarkii*) and white river crawfish (*Procambarus acutus acutus*) in cultivated, noncultivated, and wooded ponds in south Louisiana. PhD Dissertation, Louisiana State University, Baton Rouge, Louisiana, USA.

Sawyer, T. K., S. A. MacLean, J. E. Bodammer, and B. A. Hacke. 1979. Gross and microscopical observations on gills of rock crabs (*Cancer irroratus*) and lobsters (*Homarus americanus*) from nearshore waters of the eastern United States. Pages 68-91 in *Proceedings of the Second Biennial Crustacean Health Workshops*. Texas A&M University Sea Grant College Program, College Station, Texas, USA. Pbl. No. TAMU-SG-79-114.

Schaperclaus, W. 1954. *Fischkrankheiten*. Akademie-Verlag. Berlin. 108 p.

Schaperclaus, W. 1979. *Fischkrankheiten*. Akademie-Verlag. Berlin. 1089 p.

Schmidt, G. D. 1973. Resurrection of *Southwellina* Witenberg, 1932, with a description of *Southwellina dimorpha* sp. n. and a key to the genera in Polymorphidae (Acanthocephala). *The Journal of Parasitology* 59(2):299-305.

Scott, J. R., and R. L. Thune. 1986a. Bacterial flora of hemolymph from red swamp crawfish, *Procambarus clarkii* (Girard), from commercial ponds. *Aquaculture* 58:161-165.

Scott, J. R., and R. L. Thune. 1986b. Ectocommensal protozoan infestations of gills of red swamp crawfish, *Procambarus clarkii* (Girard), from commercial ponds. *Aquaculture* 55:161-164.

Sharfstein, B. A., and C. Chafin. 1979. Red swamp crayfish: short-term effects of salinity on survival and growth. *Progressive Fish-Culturist* 41:156-157.

Sheppard, M. M. 1974. Growth patterns, sex ratio and relative abundance of crayfishes in Alligator Bayou, Louisiana. Master of Science Thesis, Louisiana State University, Baton Rouge, Louisiana, USA.

Silva, J. L., E. Marroquin, and Y. M. Lai. 1991. Yield and textural properties of pond raised crayfish *Procambarus clarkii* during a harvest season. *Program and Abstracts, 22nd Annual Conference & Exposition*, World Aquaculture Society, San Juan, Puerto Rico, pp. 57-58.

Sleigh, M. 1973. *The Biology of Protozoa*. American Elsevier Publishing Company, Inc., New York, New York, USA.

Smith, R. I. 1940. Studies on two strains of *Aphanomyces laevis* found occurring as wound parasites on crayfish. *Mycologia* 32:205-213.

Smith, H. L. Jr. 1979. Serotyping of non-cholera vibrios. *Journal of Clinical Microbiology* 10:85-90.

Smith, H. L. Jr., and K. Goodner. 1965. On the classification of *Vibrio*. Pages 4-8 in O. A. Bushnell and C. S. Brookhyser (eds.), *Proceedings of the Cholerae Research Symposium*, Honolulu, Hawaii, USA.

Smitherman, R. O., J. W. Avault, Jr., L. de la Bretonne, Jr., and H. A. Loyacano. 1967. Effects of supplemental feed and fertilizer on production of red swamp crawfish, *Procambarus clarkii*, in pools and ponds. *Proceedings 21st Annual Conference Southeastern Association of Game and Fish Commissioners* 21: 452-459.

Somers, K. M., and D. M. Stechey. 1986. Variable trappability of crayfish associated with bait type, water temperature and lunar phase. *American Midland Naturalist* 116:36-44.

Sommer, T. R. 1984. The biological response of the crayfish *Procambarus clarkii* to transplantation into California rice fields. *Aquaculture* 41:373-384.

Sommer, T. R., and C. R. Goldman. 1983. The crayfish *Procambarus clarkii* from California rice fields: ecology, problems, and potential for harvest. *Freshwater Crayfish* 5:418-428.

Stechey, D. M., and K. M. Somers. 1983. An analysis of four Ontario species of crayfish for aquaculture. *Proceedings of the 1st International Conference on Warm Water Aquaculture Crustacea*, Brigham Young University, Hawaii, USA, pp. 221-230.

Stevenson, J. R. 1975. The molting cycle in the crayfish: recognizing the molting stages, effects of ecdysone, and changes during the cycle. *Freshwater Crayfish* 2:255-270.

Suko, T. 1953. Studies on the development of the crayfish. I. The development of secondary sexual characters in appendages. *Science Reports of Saitama University* (Japan) 1B:77-96.

Suko, T. 1954. Studies on the development of the crayfish. II. The development of the egg-cell before fertilization. *Science Reports of the Saitama University* (Japan). 1B:165-175.

Suko, T. 1955. Studies on the development of the crayfish. III. The development of testis before fertilization. *Science Reports of the Saitama University* (Japan) 2B:39-44.

Suko, T. 1956. Studies on the development of the crayfish. IV. The development of winter eggs. *Science Reports of the Saitama University* (Japan) 2B:213-219.

Suko, T. 1958. Studies on the development of the crayfish. VI. The reproductive cycle. *Science Reports of Saitama University* (Japan) 3B:79-91.

Suko, T. 1961. Studies on the development of the crayfish. VII. The hatching and the hatched young. *Science Reports of the Saitama University* (Japan) Series B, B(1):37-42.

Tabrosky, P. 1982. The white prince. *Tropical Fish Hobbyist* 30(82-01):8-13.

Tack, P. I. 1941. The life-history and ecology of the crayfish Cambarus immunis Hagen. *American Midland Naturalist* 25:420-446.

Taketomi, Y., M. Murata, and M. Miyawaki. 1990. Androgenic gland and secondary sexual characters in the crayfish *Procambarus clarkii*. *Journal of Crustacean Biology* 10:492-497.

Tanchotikul, U., and T. C.-Y. Hsieh. 1989. Volatile flavor components in crayfish waste. *Journal of Food Science* 54(6):1515.

Taylor, R. C. 1984. Thermal preference and temporal distribution in three crayfish species. *Journal of Comparative Biochemistry and Physiology* 77A:513-517.

Thune, R. L., J. P. Hawke, and R. J. Siebling. 1991. Vibriosis in the red swamp crawfish *Procambarus clarkii*. *Journal Aquatic Animal Health.* 3:188-191.

Trimble, W. C., and A. P. Gaudé, III. 1988. Production of red swamp crawfish in a low-maintenance hatchery. *Progressive Fish-Culturist* 50:170-173.

Turner, H. M. 1984. Orientation and pathology of *Allocorrigia filiformes* (Trematoda:Dicrocoeliidae) from the antennal glands of the crayfish *Procambarus clarkii*. *Transactions of the American Microscopical Society* 103(4):434-437.

Turner, H. M. 1985. Pathogenesis of *Alloglossoides caridicola* (Trematoda) infection in the antennal glands of the crayfish *Procambarus acutus*. *Journal of Wildlife Diseases* 21(4):459-461.

Unestam, T. 1969a. On the adaptation of *Aphanomyces astaci* as a parasite. *Physiologia Plantarium* 22:221-235.

Unestam, T. 1969b. Resistance to the crayfish plague in some American, Japanese, and European crayfishes. *Reports of the Institute of Freshwater Research*, Drottningholm, Sweden 49:202-209.

Unestam, T. 1973. Fungal diseases of crustacea. *Review of Medical and Veterinary Mycology* 8(1):1-20.

Unestam, T. 1975. Defense reactions in crayfish towards microbial parasites, a review. Pages 327-336 in J. W. Avault, Jr. (ed.), *Freshwater crayfish II*. Baton Rouge, Louisiana, USA, p. 676.

Uribeondo, J. U., and K. Söderhäll. 1992. *Procambarus clarkii* can carry the plague fungus. *Crayfish News. IAA Newsletter* 14(2):4.

Vey, A., K. Söderhäll, and R. Ajaxon. 1983. Susceptibility of *Orconectes limosus* Raff. to the crayfish plague *Aphanomyces astaci* Schikora. *Freshwater Crayfish* 5:284-291.

Viosca, P., Jr. 1937. *Pondfish culture*. Pelican Publishing Co., New Orleans, Louisiana, USA.

Viosca, P., Jr. 1966. *Crawfish farming*. Louisiana Wildlife and Fisheries Commission Wildlife Education Bulletin No. 2, New Orleans, Louisiana, USA.

Vogelbein, W., and R. L. Thune. 1988. Ultrastructural features of three ectocommensal protozoa attached to the gills of the red swamp crawfish, *Procambarus clarkii*. *Journal of Protozoology* 35(3):337-344.

Westman, K., M. Pursiainen, and P. Westman. 1990. Status of crayfish stocks, fisheries, diseases and culture in Europe. *Report of the FAO European Inland Fisheries Advisory Commission (EIFAC) Working Party on Crayfish*. Riista-ja Kalatalouden Tutkimuslaitos/Kalatutkimussia-Fishkundersokningar, Yliopistopaino, Helsinki, Finland.

Witzig, J. F., J. W. Avault, Jr., and J. V. Huner. 1983. Crawfish (*Procambarus clarkii*) growth and dispersal in a small south Louisiana pond planted with rice (*Oryza sativa*). *Freshwater Crayfish* 5:331-343.

Witzig, J. F., J. V. Huner, and J. W. Avault, Jr. 1986. Predation by dragonfly naiads *Anax junius* on young crawfish *Procambarus clarkii*. *Journal of the World Aquaculture Society* 17:58-63.

Xinya, S. 1988. Crayfish and its cultivation in China. *Freshwater Crayfish* 7: 391-395.

Young, Willard. 1966. Ecological studies of the Branchiobdellidae (Oligochaeta). *Ecology* 47(4):571-578.

Cultivation of Freshwater Crayfishes in Europe

Hans Ackefors

Department of Zoology
Stockholm University
Stockholm, Sweden

Ossi V. Lindqvist

Department of Applied Zoology
University of Kuopio
Kuopio, Finland

Section I:
Introduction

The recorded history of the use of crayfish as well as of their culture and management in Europe extends back hundreds of years, at least to the Middle Ages. In central Europe, Emperor Maximilian I of the Habsburg Empire and the Archbishops of Salzburg gave written orders on crayfish catching. Monasteries of the time in the fifteenth century had their own ways of cultivating, holding, and cooking crayfish (Spitzy 1973), apparently to be consumed especially as a lenten fare. Earlier still, in the old Roman Empire, crayfish appear in a recipe from the third century A.D. published by Caelius in *De re culinaria* (Lagerqvist and Nathorst-Böös 1980).

In Scandinavia, written records on the interest of the Swedish royal house in crayfish come from the sixteenth century (Abrahamsson 1969). The Wasa Kings introduced the noble crayfish *Astacus astacus* from Germany in the early sixteenth century and kept them in ponds. Most likely natural crayfish populations existed

157

prior to this introduction. The wide distribution of crayfish in southern Sweden indicates that *Astacus* is native in Sweden (Abrahamsson 1972), and also in southern Finland (cf. Westman 1973a).

The so-called noble crayfish, *Astacus astacus* L., was the delicacy of the noblesse but gradually gained popularity among the commoners. In the mid-nineteenth century especially in Paris and St. Petersburg, large numbers of crayfish were sold (cf. Kusnetzow 1898). In the period between 1853 and 1879 more than 5 million crayfish were annually consumed in Paris alone. Most of this trade came from Germany and Russia (Smolian 1926). Thus the crayfish recipes spread mostly with the French cuisine. This popularity may be one of the reasons why crayfish also attracted scientific interest in the nineteenth century. The classical work on crayfish biology by T. H. Huxley (Huxley 1880), and the earlier studies on crayfish tissues by the famous Ernst Haeckel (Haeckel 1857) are two prime examples.

A new interest in crayfish management and cultivation arose in the latter half of the nineteenth century and early part of the twentieth century, for a multitude of reasons. Expanding industrialism created polluted waters and waterways. Acidification had become a problem even then (cf. Nolte 1933). The demand for crayfish apparently created "overfishing" situations in central Europe and elsewhere. However, the overriding problem was the appearance of the "crayfish plague" and its devastating effect on native European crayfish stocks in the late nineteenth and early twentieth century.

Finland and Sweden may provide examples of the overall effects of the crayfish plague fungus *Aphanomyces astaci*. The total crayfish (*Astacus astacus*) production in Finland in some years at the turn of the century was estimated to be as high as 700 tons with over 500 tons being exported (Nordqvist 1898; Järvi 1910), while the annual yield in the 1980s has been of the order of 100 tons. Similarly, the yield of the noble crayfish in Sweden before the plague was about 1000 tons annually, while in the 1980s it hardly reached 100 tons and may be now down to 50 tons annually (Fürst 1990). Care should also be exercised in the examination of these comparisons, because the yields do not directly indicate biological production without reference to the fishing effort and methods.

The earliest attempts at crayfish cultivation have been referred to by several workers including Smolian (1926) in Germany, Laurent (1973) in France, and Sestokas and Cukerzis (1973) for Lithuania. These early attempts were mostly directed towards restocking former crayfish waters that had become depleted. Another step towards restocking was the introduction of a (plague-resistant) American crayfish species *Cambarus affinis* Say (= *Orconectes limosus* (Raf.)) to Germany in 1890. *Orconectes limosus* is smaller in size than *A. astacus* and thrives better in eutrophic waters. Although abundant in places in Western Europe, *O. limosus* commands little, if any, commercial interest nowadays.

The new wave of interest in crayfish cultivation began in Europe in the 1960s and 1970s with the introduction of the North American astacid, *Pacifastacus leniusculus* Dana, the signal crayfish (Ackefors et al. 1989). It was further enhanced by the formation of the International Association of Astacology in 1972. The Symposium publications, titled *Freshwater Crayfish, A Journal of Astacology,* have been issued in seven volumes up to 1990.

Today crayfish are consumed mostly as a luxury food item, mostly in Sweden and France, but also in Germany, Spain, Italy, and Finland. Cultured crayfish are also finding new customers in England. In most other European countries there is little interest in eating crayfish. Yet crayfish catching is common in many countries with productive stocks though the local people themselves may not consume them. The main producers of crayfish for the European market are the former Soviet Union, Spain, the USA, Turkey, and also the People's Republic of China. Turkey lost much of its production since the mid-1980s because of the arrival of the crayfish plague fungus. The total consumption of crayfish in Europe, before the collapse of the Turkish production in 1985-1987, may have been close to 10,000 tons. According to a recent estimate some 8000-8500 tons of freshwater crayfish are produced annually (Westman et al. 1990). With importation from countries outside Europe the consumption may exceed 10,000 tons.

Section II:
The Crayfish and Crayfish Production in Europe

There are five recognized native and three introduced crayfish species in Europe. All the native European species belong to the family Astacidae; the North American *Pacifastacus* is also a member of Astacidae. There is still some confusion about the taxonomic status of many local "phenotypes," as well as the separation of the genera, e.g., *Astacus*, from *Austropotamobius*. Laurent (1988) gave a summary of the observations and studies of interactions between European species in overlapping ecological situations and noted the absence of successful hybridization between them, including also *Pacifastacus leniusculus*.

Astacus astacus L., the noble crayfish, is also called the broad-clawed crayfish (Figure 52). This species used to be the main target of crayfish trapping in Europe, but its stocks have been hardest hit by the crayfish plague fungus *Aphanomyces astaci* and also by water pollution and habitat changes.

The present distribution of *A. astacus* has contracted from that occurring earlier, especially in central Europe. In France, for instance, the species now occurs only in scattered places in the north and northeast of the country (Laurent and Forest 1979). Previously, the species had wide distribution throughout eastern Europe (e.g., Herfort-Michieli 1973) and western Ukraine and the Baltic states (Cukerzis 1984) south to parts of Greece.

The stocks of *A. astacus* are more scattered now. Main production areas nowadays are the Nordic countries, Finland, Sweden, Norway, and Denmark, with more patchy production in central, eastern, and southern Europe. In Finland, the original northern range of *A. astacus* extended to 62°N and in Sweden to 61°N, but introductions through this century have expanded its distribution in Finland and Sweden, for example, even north of the Arctic Circle (Westman 1973a). Breeding is generally confined to waters where temperatures during three summer months exceed 15°C (Abrahamsson 1972).

The annual catches in Finland and Sweden of *A. astacus* are

Figure 52. Mature male noble crayfish, *Astacus astacus.* O. Lindqvist.

estimated at 100 and 50 tons, respectively, but it is much smaller in
Norway and Denmark. The legal size was 10 cm total length (TL) in
Finland but has been eliminated altogether in 1993. It is 9 cm TL in
Sweden. In Finland, the trapping season opens July 21st, and the
second Wednesday of August in Sweden. The season lasts until the
end of October in Finland and until the end of December in Sweden.

Astacus leptodactylus Esch. is the Turkish or narrow-clawed
crayfish (Figure 53). The main area of distribution extends east of
A. astacus, but there is some overlapping in the Baltic area and
Poland (cf. Köksal 1988). In many lakes in Turkey it formed dense
and productive populations with annual yields in the early 1980s of
some 7000 tons that were the basis of extensive exports to western
and northern Europe before the plague fungus arrived (see Erencin
and Köksal 1977; Huner 1984; Köksal 1988). The taste of this
species does not compare with that of the noble crayfish. The meat
yield is also lower because there is less muscle in the narrow claws.
The species reaches about the same size as the noble crayfish, i.e.,
8-12 cm TL on the average. It has been introduced in scattered
places in western Europe and has been an object of aquaculture
(e.g., Vigneux 1979).

Austropotamobius torrentium Schrank is the stone crayfish. This
species is of minor commercial importance. It occurs in clear waters
in mountainous areas in central and southeastern Europe, e.g., in
Switzerland, Austria, and Yugoslavia (cf. Wintersteiger 1985; Laurent 1988). It thrives in cooler waters than *A. pallipes* or *A. astacus*
and remains smaller in size.

Austropotamobius pallipes (Lereboullet) is the white-footed crayfish (Figure 54). It occurs in western Europe west of the Rhine
River, in France, the Iberian Peninsula, and Italy (Laurent and Forest 1979) as well as England, Wales, and Ireland (Moriarty 1973;
Jay and Holdich 1981; Reynolds 1988). In the British Isles it has
not been exploited to any great extent (Rhodes and Holdich 1979),
but in Spain the natural populations have been used with variable
yields (Huner 1984). In France the exploitation of this species for
human consumption has come to a virtual stop (P. Laurent, personal
communication). In Spain it is being actively cultivated on a limited
basis (Habsburgo-Lorena 1983a). The *A. pallipes* stocks in the Brit-

Figure 53. Mature male narrow-clawed or Turkish crayfish, *Astacus leptodactylus.* D. Holdich.

Figure 54. Mature male *Austropotamobius pallipes.* D. Holdich.

ish Isles have been hit by the crayfish plague fungus (Alderman et al. 1984), apparently as result of the introduction of the signal crayfish (Goddard and Hogger 1986; Marren 1986; Holdich et al. 1990).

The fifth native European species is *Astacus pachypus* Rathke. This species occurs in the south of what previously was the Soviet Union, in areas next to Black Sea and Caspian Sea (cf. Karaman 1962; Brodsky 1983). Little is known about it.

The introduced species are as follows:

Orconectes limosus (Raf.), the striped crayfish, is a cambarid species that was introduced into River Metzer, Germany from Pennsylvania in 1890. It is now widely spread especially along the big rivers and waterways in Central Europe including Germany, the Netherlands, Poland, and France (Laurent and Suscillon 1962; Kossakowski 1973; Schweng 1973), but because of its small size, it commands little commercial interest. In fact, in some places it is considered to be a nuisance.

Orconectes virilis was introduced into France in 1897 and in Sweden in 1960. No self-sustaining populations developed (Lowery and Holdich 1988).

Pacifastacus leniusculus Dana is the signal crayfish (Figure 55). This species was the second one considered for introduction to Europe after *O. limosus*. The purpose of the introductions was to offset the losses of native crayfish to the crayfish plague fungus. The introduction was preceded by ecological studies (e.g., Abrahamsson and Goldman 1970). The stockings took place in Sweden initially, beginning in 1960 with adult signal crayfish from California and later from British Columbia (Svärdson 1965; Abrahamsson 1973b; Fürst 1977). In 1970-71 some adult and subadult signal crayfish were also introduced into Austria from California. In Finland, adult signal crayfish were imported from California in 1967-68 and were subsequently supplemented with juveniles from a Swedish hatchery (cf. Westman 1973a). It is important to remember that this species of crayfish carries the crayfish plague fungus *A. astaci* in their exoskeleton as a latent infection (Persson and Söderhäll 1983). It can thus function as a vector for this disease for susceptible crayfish species.

After about 1971 no other known direct introductions have been

Figure 55. Mature male signal crayfish, *Pacifastacus leniusculus.* D. Holdich.

made from North America and the interest in local cultivation of *Pacifastacus* has increased in several countries. A great majority of all stocking material in Europe has been produced by a private hatchery in Blentarp, Sweden. In 1970-72 alone, the production was some 0.5 million juveniles (Abrahamsson 1973b; Brinck 1983). Elsewhere in Europe the signal crayfish has been introduced to well over a dozen countries, including England, Germany, France, Spain, Poland, and the former Soviet Union. In Spain, for instance, the first official introduction took place in 1976, and the stocks seem to be on the increase, especially in cool waters (Habsburgo-Lorena 1979, 1983a).

The reasons for and success of introductions of the signal crayfish are related to its apparent relative resistance against the plague fungus (Unestam 1973), its higher growth rate and consequent earlier maturity compared to *A. astacus* (Abrahamsson 1973a, b), its apparent ecological success in crayfish waters especially in Sweden (Brinck 1977, 1983), and its acceptance by consumers, as its meat is said to be comparable to that of the noble crayfish. (Yet its shell is harder to break than that of *A. astacus*.) Furthermore, its qualities may make it suitable for aquaculture as it apparently responds slightly better to feeding and tolerates somewhat higher temperatures than *A. astacus* (cf. Mason 1975; Strempel 1975; Laurent et al. 1979). In contrast, the signal crayfish usually harbor the crayfish plague fungus as a chronic infection in the cuticle and may then carry on an infection to non-resistant stocks (Unestam 1973; Unestam et al. 1977). Under stress the signal crayfish may also succumb to the plague fungus. This has to be considered an additional risk factor in the culture. *Orconectes limosus* may also carry the fungus as a chronic infection (Vey et al. 1983; Persson et al. 1987; Söderhäll 1988). Resistance to the plague fungus is a powerful "weapon" in an ecological competition between the species.

Procambarus clarkii (Girard) is the red swamp crayfish (Figure 5). So far, this warm-water species from the southern USA has been successfully introduced in Spain and Portugal (Habsburgo-Lorena 1979, Ramos and Pereira 1981, de Bikuna et al. 1989). It is also found in the wild in the south of France (Laurent et al. 1991). It has been cultivated experimentally in Italy (Mancini 1986). Some specimens have been reported from England, the Netherlands, and even

Sweden (Huner 1988), indicating that this species can persist though not necessarily breed in those environments. More detailed information about the biology and exploitation of *P. clarkii* can be found in the chapter on the culture of *P. clarkii* and in the North American section of this book.

In Europe, this species is most abundant in Spain where the first introductions took place in 1972-73, in the southern part of the country. The red swamp crayfish inhabit paddy fields and slow-moving warm waters and is still expanding its range. The first catches in 1976 were estimated at less than a ton, but by 1980 they had increased to 350 tons (cf. Habsburgo-Lorena 1983a,b; Huner 1984). In 1986 and 1987 the catches were 3384 and 4650 tons, respectively (Habsburgo-Lorena 1990). Some 400 full-time fishermen obtain their living in Spain from trapping *P. clarkii*, and the main seasons are from March to May and from September to October (Habsburgo-Lorena 1983a,b). Spain is currently the major producer of crayfish in Europe, though it exports little (Westman et al. 1990).

The production of *P. clarkii* in Spain is still expanding, in contrast to the diminishing stocks of *A. pallipes* and *A. leptodactylus* which are suffering from both water pollution and the spread of the crayfish plague fungus (Cuellar and Coll 1983; Morales 1987). At least two hatcheries produce the young of *P. clarkii* for stocking purposes, but the crayfish are also managed in places by additional feeding in ponds and raceways (Habsburgo-Lorena 1983a,b). France bans the importation of live *P. clarkii* (Westman et al. 1990).

Turkey has been, historically, the leading European country in crayfish production from her lakes. The yield in the early 1980s reached 6000 to 7000 tons annually; all of this was exported. The only species harvested at present is *A. leptodactylus* but after 1984 the crayfish plague fungus infested most of the productive waters (Fürst and Söderhäll 1987; Rahe and Soylu 1989) reducing yield by at least 85%.

The natural fisheries for *A. astacus* in the Nordic countries of Finland, Sweden, and Norway produce a total in excess of 200 tons annually. Most of this harvest is consumed locally with the exception of Norway where a good part of the yield is exported to the neighboring Sweden. Some 300 tons of the introduced *P. leniuscu-*

lus were caught in 1989 (Svärdson et al. 1991), however, only 50 tons of *A. astacus* were caught. The present establishment of culture units indicate that *P. leniusculus* will dominate the future harvests in Sweden (Ackefors 1991). The Finnish catch (of *A. astacus*) has been rather steady at 100 tons annually. Although introduced since the late 1960s, few *P. leniusculus* are harvested in Finland and populations are sparse (Westman et al. 1990).

France and Sweden were the major importing countries of crayfish in Europe in the early 1980s. Both of them imported some 2000 tons annually. Most came from Turkey, but since 1987 onwards most of the crayfish have come from the USA. This is primarily *P. clarkii* from Louisiana but a modest amount of *P. leniusculus* is imported from the West Coast of the USA (Thompson 1990). According to the Food and Agriculture Organization (FAO) of the United Nations statistics this has changed during recent years (Table 20). Italy and Sweden are now the greatest importers of freshwater crayfish. Italy imported 2170 tons and Sweden 1943 tons in 1990. In total 429 tons of crayfish were imported to various countries in Europe to a total value of 45 million US dollars.

Spain is now the greatest European exporter in terms of quantity. In 1988 the country exported 357 tons and in 1990 867 tons according to preliminary statistics. The value of the export was 3.9 million US dollars. The net import to Europe was 3800 tons and the trade deficit was 34 million US dollars.

The total production (and consumption) of crayfish in Europe may well have been in excess of 7000 tons annually (cf. Huner 1984 who cites an earlier figure of 6000 tons), but this was before the collapse of the Turkish fishery. With the increase in the Spanish production and increased imports, the annual European consumption of crayfish stands at between 8000 and 10,000 tons.

The total annual production in Europe may now be on the order of 7000-8000 tons (Laubier and Lindqvist 1990; Westman et al. 1990), but great annual variations may still exist. Virtually all of the European production comes from natural fisheries, and aquaculture production for human consumption amounts to a few tons only (in the United Kingdom) (Huner 1984) and less than 10 tons in Sweden (Ackefors 1991). However, part of the production of *P. clarkii* in

Spain comes from a kind of extensive management of the stocks in rice fields, ponds, and raceways.

The trade of crayfish within Europe has undergone rapid changes in response to the production patterns in different countries. Table 20 gives a summary of the imports and exports of crayfish in Europe in 1990, according to the statistics provided by FAO. The relatively high market value of freshwater crayfish in Europe makes this commodity an important commercial item. The recent general increase in prices has also stimulated interest in the culture of crayfish, even in countries with little consumption of their own (cf. Laubier and Lindqvist 1990).

Table 20. The import (I) and export (E) of freshwater crayfish by various European countries in 1990, in metric tons and valued as thousands of $US.

Country	Import		Export		Balance	(E-I)
	Tons	Value	Tons	Value	Tons	Value
Belgium	163	1311	62	400	−101	−911
Denmark	79	411	0	0	−79	−411
Finland	77	745	1	7	−76	−738
France	748	3748	3	20	−745	−3728
Germany	72	679	0	5	−72	−674
Greece	0	0	18	374	+18	+374
Italy	2170	21616	514	4454	−1656	−17162
Netherlands	140	795	34	406	−106	−389
Portugal	0	1	70	377	+70	+376
Spain	34	69	867	3860	+833	+3791
Sweden	1943	15434	45	265	−1898	−15169
U.K.	3	41	48	878	+45	+837

The European 'imbalance' is −3767 tons, valued at US$30 mill.

Section III:
Crayfish Culture

GENERAL

The noble crayfish *Astacus astacus* was originally abundant in many European countries. As a food it was considered superior to all other crayfish species in Europe. Great quantities were exported to Germany, France, and Russia at the end of the nineteenth century. Currently the stocks of *A. astacus* in most European waters have declined sharply, for various reasons, the greatest of which is the crayfish plague fungus (cf. Lindqvist 1988). The crayfish culture situation in Europe during the last 30 years has been highly varied. Many of the early cultural attempts have failed for economic and other reasons. The choice of species is important for both biological and commercial reasons. The species must tolerate and thrive under local environmental conditions. Risks should be manageable and the product should have steady markets as the production cycle in temperate and cool regions usually takes several years. Crayfish culture takes two forms. Early or one-summer juveniles are produced for restoration of natural crayfish waters, and marketable-sized crayfish, generally 9-10 cm TL or even bigger, are grown within the culture facility.

The interest in crayfish culture is emphasized by the publication of several books on the subject in various languages. Arrignon (1979) provides a very detailed crayfish culture text in French. The second edition of this book is thoroughly updated and provides much useful information about *P. clarkii* (Arrignon 1990). An English translation is in preparation. Auvergne (1982) has written a booklet about enhancement of crayfish stocks in Spain. Westman and Nylund (1984) emphasize natural history and biology of Finnish crayfish in their book but also include information on crayfish cultivation. Groves (1985) has published a brief text in English on crayfish biology and culture. Mancini (1986) has written an Italian language text similar to that of Westman and Nylund in that it contains much information about natural history and biology. It was also the first European text to provide information about *P. clarkii* although the species is present in only isolated locations in Italy. This was followed by Morales' (1987) Spanish language text that emphasizes *P. clarkii.* The short French text by Arrignon et al. (1990) discusses the status of *P. clarkii* in Europe at great length. Other texts emphasize native species including *A. astacus, A. leptodactylus,* and *A. pallipes,* and the introduced *P. leniusculus* probably because the climate is much colder in those countries. Hammarlund and Karlsson (1985) and Fürst (1986) have written Swedish language booklets explaining cultural practices that have been developed in Sweden over the past two decades. The Holdich and Lowery (1988) text is an excellent English language treatment of crayfish biology, ecology, and culture. An extensive statistical type of coverage of European crayfish and their exploitation and main culture methods is given by Westman et al. (1990). Westman et al. (1992) have recently published a book on natural history and biology of Swedish and Finnish crayfish conditions including an extensive section on recent progress on crayfish farming.

THE LIFE CYCLE OF ASTACUS ASTACUS

A general outline of the breeding cycle of crayfishes of the family Astacidae, i.e., *Astacus, Austropotamobius,* and *Pacifastacus* follows. All are rather similar and best illustrated by the following description of the life cycle of *A. astacus,* as it occurs in Sweden or Finland (Figure 56).

The females of *A. astacus* become sexually mature at a total length of 7-8 cm. Mating occurs in the autumn, with declining temperatures at about 5°C. At mating, the male deposits the spermatophores on the ventral side of the female, usually close to her genital openings, but sometimes even on the tail. After a short period lasting one to several days, the female will lie on her side or back and extrude the eggs together with a fluid that dissolves the jellified spermatophore. The sperm cells are liberated and the eggs fertilized. The fluid forms a membranous substance which covers the eggs and prevents them from sticking together. The eggs are

Figure 56. The life cycle of *Astacus astacus* as it occurs in Sweden and Finland in natural-ambient water. By using heated water (20° C) after a short period in cold water the rate of the development of the eggs can be speeded up in indoor facilities (Cukerzis 1979). This procedure can increase the length of time for growth between hatching and the first winter. After Ackefors (1989).

BREEDING CYCLE OF
Astacus astacus

1. NATURAL-AMBIENT WATER

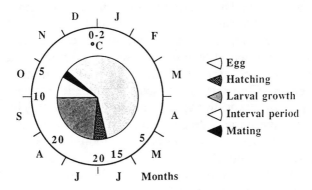

2. ARTIFICIAL-HEATED WATER

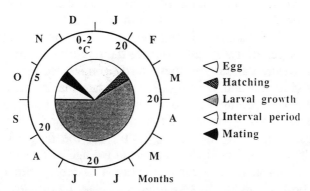

attached on the hairy swimmerets (pleiopods) under the tail. In the
north, the incubation period lasts from October for about 7-8 months
until the following June. The eggs hatch at temperatures of 17-20°C
(Cukerzis 1973). The incubation time has some relation to the de-
gree-days, and will thus vary according to the temperature.

The number of pleiopodal eggs is 50-200, correlated with the
size of the female. The number of eggs actually hatched is usually
lower as some eggs remain unfertilized or are lost for other reasons
during the incubation period. The signal crayfish *Pacifastacus leniusculus* has smaller eggs than *A. astacus,* but their numbers per
female are also higher, 300-400 on average.

After hatching, the larvae cling together under the female. After
the first molt (8 days after hatching), the larvae start to disperse
away from the female, though they may seek shelter under the tail
of the female for a period of some two weeks. During this time the
female does not feed. After 10-12 days from the first molt, when the
young are called "2nd stage juveniles," the female may start prey-
ing upon its own offspring (Munkhammar et al. 1989). The young
seek shelter in the immediate environment and molt 5-6 times during their first summer. No molts occur any more in late September,
at temperatures below 10°C (Ackefors et al. 1989). The young
overwinter as 20 mm long juveniles. In the wild, 3-6 years are
needed to grow to the commercial size of 9 cm TL in Sweden.

In southern Sweden, Abrahamsson (1971) found that *Astacus*
juveniles reached total lengths of approximately 23 mm after the
first summer, 49 mm (males) or 47 mm (females) after the second
summer, and 72 (males) or 70 mm (females) after the third summer.
The females generally matured during their fourth summer, while
the males matured already in their third summer. However, recent
results by Gydemo and Westin (1989) indicate that *A. astacus* has
the potential of reaching 9 cm total length already in three summers
(that is, at the age of 2+ years) on the island of Gotland in the Baltic.
Recent results indicate that it is possible for the fastest growing
individuals to reach 9 cm in 2 summers at the age of 1+ summers.

Sexual dimorphism is clear in mature *A. astacus* with females
having a broader abdomen and smaller chelae than males (Figure
57). This phenomenon is well described (Lindqvist and Lahti 1983;
Huner et al. 1991).

Figure 57. Mature noble crayfish, female, left, and male, right. Note the enlarged abdomen of the female. J. Huner.

CULTURE OF ASTACUS ASTACUS

Artificial Incubation of Eggs

Fertilized and stripped eggs can be incubated in trays or in incubation flasks (Cukerzis 1973, Cukerzis 1979, Järvenpää 1987). Cukerzis (1988) described in detail the construction of a one-liter incubation flask and the procedures for hatching eggs.

Järvenpää and his colleagues in Finland have tested the use of a mist chamber for egg incubation. The method as such works well, although it takes too much labor to remove nonviable eggs before they contaminate others. The eggs have to be returned to water before hatching.

Later, Järvenpää and Ilmarinen (1990) tested the use of a moving tray for artificial incubation of crayfish eggs. The survival of stripped eggs in moving trays was found to be better than that on standing trays. This method also saves labor as it virtually eliminates the need to remove dead eggs.

An additional advantage of artificial incubation is the saving of space compared to the rearing of berried females. The possible transmission of parasites or diseases from mother to the offspring can be prevented by proper chemical treatment of the stripped eggs. Some disease agents like the one causing *Thelohania* might be transferred via ovaries (Rolf Gydemo, Askö Laboratory, University of Stockholm, Stockholm, Sweden, personal communication).

One of the pioneer crayfish culturists has been J. Cukerzis with his team in Lithuania (Cukerzis 1959). His early experiments were aimed at producing young of *Astacus astacus* for restocking in natural waters. He used small ponds (3.5-5.0 m in diameter) for this production. The parent crayfish were caught in the wild. He either caught berried females early in the summer and incubated them with their eggs attached, or he kept the parent animals in ponds or containers through the winter so that copulation and incubation took place under controlled conditions. The young were stocked in ponds after hatching and harvested late in the summer for transfer to lakes and rivers. The same basic culture method has been widely adopted, with modifications and improvements in other countries. Summaries of these methods are described by Cukerzis (1979, 1984, 1988). The latest article is a revised form of an earlier 1970 text that was translated into French. Although rarely practiced, incubation of eggs in regular upright glass incubators has been successful after stripping them from the female. The stripped eggs may be prone to damage if mishandled, though their treatment is consequently much easier. (Examples of pond culture systems used in Finland are shown in Figures 58, 59, 60, and 61).

There are thus at least three different ways of producing crayfish juveniles:

1. The females are kept in small tanks with horizontal net partitions, one female per tank. The second stage juvenile (after the first molt) can pass through the net and thus escape the possible predation by the female.

2. The females are individually held in tanks and removed after the juveniles reach the second juvenile stage.
3. The eggs on the tail of the female are stripped off with forceps and then brought to an incubator which may also be a mist chamber.

Cukerzis (1979) demonstrated the possibility of shortening the incubation period. The total time of incubation can be decreased from the usual 7-8 months (in Lithuania) to 3-4 months by using heated water in indoor basins (cf. Figure 56). However, the eggs require a cold, quiescent period in winter for proper development, a requirement that is somewhat analogous to the incubation of eggs of some coldwater fish, especially salmonids. Hatching of juveniles can therefore occur at the end of February (at 34-46 mg hatching weight). Weights of 120-900 mg at the end of June and 200-1500 mg at the end of August can be achieved. In central Sweden, the

Figure 58. Fish hatchery ponds typical of those used for outdoor crayfish breeding in Finland. J. Huner.

Figure 59. Crayfish nursery ponds at Evo, Finland fish hatchery. J. Huner.

juveniles weigh only 140-220 mg under natural conditions in September-October. Great variation in individual weights occur in crayfish grown under culture conditions.

It is therefore possible to reduce the incubation time to hatching by slowly raising the temperature from the end of December on. The eggs can be made to hatch at the end of February at the earliest or at any time up to June-July depending on the temperature of the water. This accelerated development will result in juveniles reaching a weight of some 600 mg by the end of August and 1400 mg by January (Cukerzis 1979).

A still more rapid growth rate was demonstrated by Gydemo and Westin (1989). The larvae of *A. astacus* reached a total length of 40 mm and a weight of 2000 mg in six months (Table 21). The young crayfish were fed frozen *Bosmids*. Accordingly, the density of juveniles should not exceed 50-100 per square meter (Ackefors et al. 1989; Gydemo and Westin 1989).

Figure 60. Crayfish nursery pond at Evo, Finland with a plastic cover to heat water. J. Huner.

Pond culture of crayfishes (both *A. astacus* and *P. leniusculus*) in Sweden involves both extensive culture in ponds and semi-intensive culture in raceway-like ponds (Fürst 1986; Hammarlund and Karlsson 1992). The extensive ponds are modified to prevent fish from entering through the pond drain, and the water depth is adequate to ensure that the crayfish population will survive the winter. The semi-intensive culture ponds are narrow, 5 m or so in width, and shallow, 1.5-2.25 m in depth. Fürst (1986) distinguishes between winter ponds that are much shallower, 0.5-1.0 m deep. The summer ponds are unsuitable for keeping crayfish over the long, cold Scandinavian winters because they may freeze solid.

Semi-intensive ponds are designed to have continuous water flow and reported production has reached 1 kg/m of shore line. Production in extensive culture ponds is much lower. However, the value of crayfish still makes the extensive cultivation of crayfish in

Figure 61. Ponds and small lakes such as this are excellent places for establishing sustaining populations of noble crayfish or signal crayfish in Scandinavia. J. Huner.

Sweden a worthwhile endeavor. It takes several years for sustaining crayfish populations to become established in the two types of ponds. Some supplemental feeding may be practiced in the semi-intensive culture ponds. The total number of licenses for establishments in crayfish farming in Sweden was about 450 in 1989 (Ackefors 1991).

Extensive pond culture of *A. astacus* is developing widely on the Swedish island of Gotland where farmers are stocking irrigation ponds with crayfish (Rolf Gydemo, Askö Laboratory, Institute of Systems Ecology, Stockholm University, Stockholm, Sweden, personal communication 1991). These crayfish reach the legal size of 9.0 cm after 3 summers. Noble crayfish are very high priced in Sweden so this form of aquaculture is profitable with little input except for stocking material and some harvesting costs. The ponds are excavated with islands or peninsulas left in the center to in-

Table 21. The average weight and length changes of juvenile *Astacus astacus* at constant temperature of 20° C in Gotland, Sweden. Initial weights July 15, 1985 taken from Ackefors et al. (1989), and other values from Gydemo and Westin (1989). Both sexes pooled.

Date	Age	Initial density 64/m²		Initial density 349/m²	
	months	Length mm	Weight mg	Length mm	Weight mg
Aug. 1	1	12	64	12	64
Nov. 1	4	27	487	25	350
Feb. 1	7	41	2099	36	1389
May 1	10	48	3729	42	2393

crease the littoral area which is the highest quality crayfish habitat. Size rarely exceeds 1 ha.

In Norway, *A. astacus* is the only species allowed by law and the government has taken strict measures to keep the crayfish plague fungus out of Norwegian waters. Norway was free of the disease during the 1980s, though it has appeared now in a watercourse next to the Swedish border (Westman et al. 1990). The Norwegians are developing crayfish culture (Skurdal and Hessen 1985; Hessen et al. 1987), which, for instance, may utilize heated waters through heat exchangers from hydroelectric facilities. The number of active crayfish farms or licenses issued numbers about two dozen in Norway.

Pond culture of crayfishes in northern Europe is constrained by the fact that it takes 5-10 years to establish sustaining populations of crayfish (Huner and Lindqvist 1986), though the time may be shortened considerably by supplemental feeding and other measures. Several different age classes of reproductive adults are necessary to sustain annual cropping and it takes 2-4 years, depending on latitude, for the originally stocked juveniles to reach maturity and spawn and an equivalent time for the offspring to mature and spawn. Adult crayfish are often stocked to shorten this time constraint. Fortunately, the prices for live, domestically produced crayfish, often exceeding US$30-$100 per kg, makes it attractive for pond owners to manage smaller natural ponds, construct small ponds, or engage in more intense pond culture with the raceway-like ponds described by Hammarlund and Karlsson (1992) and Arrignon (1979).

Soft-shelled crayfish are not currently produced now in Europe, though some research attempts have been made in this direction using *A. astacus* in Finland (Huner and Lindqvist 1984).

Production of Two Broods the Same Year

By manipulating the water temperature, it is possible to induce reproduction of *Astacus astacus* during seasons other than the autumn. By shortening the annual temperature cycle, simulated in indoor basins, two growth (molt) periods and two mating periods can be produced within the span of one year (Westin and Gydemo 1986). More juveniles can then be produced within a year. Mating in indoor basins was induced by dropping the water temperature down to 5°C. By raising the temperature to 20°C, molting was induced. Under these conditions temperature appeared to be the principal factor controlling both spawning and molting.

CULTURE OF PACIFASTACUS LENIUSCULUS

The new surge in interest in crayfish culture in Europe has developed due to the arrival of the signal crayfish, *P. leniusculus*. The leading producer of the young of this species has been the Simontorp Laboratory in Sweden where production takes place entirely indoors using recirculating water (Karlsson 1977). The parent crayfish are collected from lakes and rivers in late fall. They are checked for internal ovarian development and the condition of the cement glands. Breeding occurs in large concrete raceways which are provided with roofing tiles for cover. Females with eggs are transferred to smaller troughs which are provided with partitions towards the end of the incubation period so that each female is housed individually. The water is recirculated through a filter. The optimum temperature is 13-14°C, but the temperature can also be regulated so that free-living young can be produced at the proper time for direct spring stockings. The crayfish in the troughs are fed boiled, peeled potatoes. Fish flesh spoils the water quickly and is not used. The current price is about US$0.40 per juvenile (= third development stage).

In Finland, the first experiments on crayfish culture were conducted by Westman (1973b) with *P. leniusculus*. He found that the optimum temperature for juveniles larger than 2.5 cm TL was 13-16°C, though results from more recent experiments suggest this temperature range may be too low (cf. Cedrins 1985) also depending on season. Cannibalism was found to be a major factor in mortality.

Some details of Finnish crayfish culture methods follow as based on reports of Pursiainen and Järvenpää (1981) and Pursiainen et al. (1983). The parent crayfish are trapped in August when trapping is generally still successful and held in earthen ponds with a stocking density of 6 per square meter. The sex ratio is 3:1 between females and males. The crayfish obtain natural food which is supplemented with pieces of fish meat and alder leaves. The ponds are emptied the following June with survival of over 90% and most of the females will be berried. The females are then held in special hatching boxes with upper and lower compartments separated by a coarse mesh floor. The young crayfish hatch and fall through the floor to the lower compartment when they become free living in the third development stage (Figure 62).

The free-living young are transferred into earthen ponds or fiberglass containers which are both provided with roofing tiles and such for hiding places. In the earthen ponds, the young feed on natural food. Some mortality occurs from predatory insects. However, survival in fiberglass containers is lower, perhaps because of nutritional problems as all food, plant material, fish feeds, and fish flesh must be provided. The survival rate in earthen ponds has averaged 67% with an initial stocking density of 100 per square meter. The weight of the crayfish at the end of the summer (October) ranges from 160-220 mg depending on initial densities. The maximum rearing temperature during the summer has been 22°C.

A modification of the original incubation system has been described by Pursiainen and Saarela (1985). Individual females are placed inside a horizontal plastic drainage tube. A battery of these tubes is held together by a wire mesh container. There are several vertical holes in the wall of each tube. Water flows horizontally through each tube. This makes water exchange much more efficient and eliminates problems of low oxygen sometimes experienced with trays with individual, vertical compartments separated by partitions. In addition, it is much more convenient to handle these incubation units.

Another modification is presented by Järvenpää (1987) in Finland. In winter, during the month of March, the eggs are removed

Figure 62. Production methods of *Pacifastacus leniusculus,* Method 1. Finnish crayfish culture method for raising *Pacifastacus leniusculus* according to Pursiainen and Järvenpää (1981) and Pursiainen et al. (1983). Mature crayfish are stocked with a sex ratio of 1:3, male to female. After Ackefors (1989).

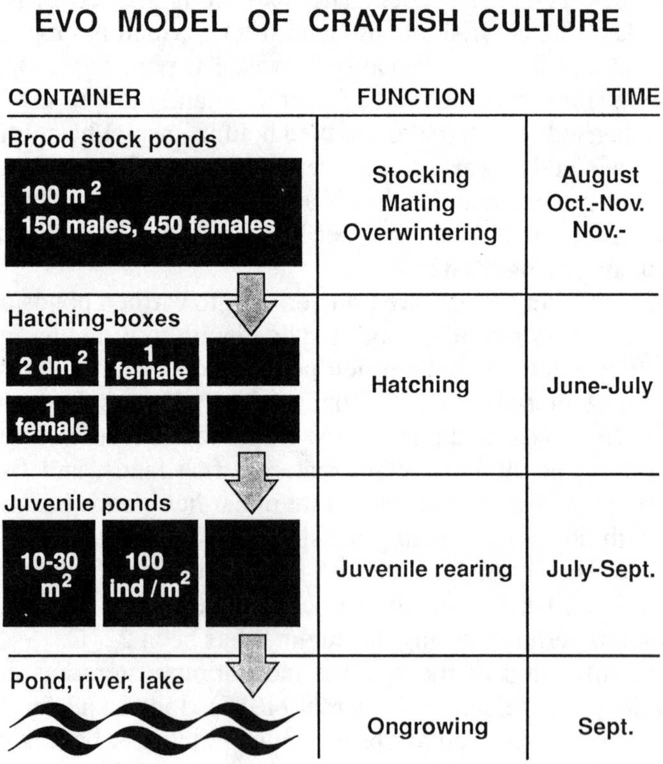

EVO MODEL OF CRAYFISH CULTURE

CONTAINER	FUNCTION	TIME
Brood stock ponds 100 m^2 150 males, 450 females	Stocking Mating Overwintering	August Oct.-Nov. Nov.-
Hatching-boxes 2 dm^2 / 1 female / 1 female	Hatching	June-July
Juvenile ponds 10-30 m^2 / 100 ind /m^2	Juvenile rearing	July-Sept.
Pond, river, lake	Ongrowing	Sept.

from their mother and placed on incubation trays in running well water (Figure 63). In April, the water temperature is gradually raised to 13-15°C. Fungal infections are treated with malachite green twice a week. The first juveniles hatch after a month in early May, nearly 2 months earlier than in the wild. In ponds, these early juveniles reach a length of 4-5 cm by September. By the end of the second summer they have grown to 7-8 cm, and at least half of the population is sexually mature.

Juveniles can also be raised in indoor tanks with heated water throughout the winter (Westman and Nylund 1984). Under these conditions the optimal temperature for growth was found to be 21°C, though overall survival is better at lower temperatures. The individuals exhibiting the maximum growth can reach 10 cm total length in 12 months.

The number of farms or facilities involved in culture of crayfish, both *A. astacus* and *P. leniusculus,* is 40-50 in Finland. They mostly produce one-summer juveniles for stocking purposes, though the profitability of the operations has been variable at best, due to the rather limited demand.

Considering the variety of climates and growing seasons in Finland, profitable culture of crayfish (*A. astacus* at least, but also *P. leniusculus*) through all life stages to the market size (about 10 cm TL) is possible only at elevated temperatures. In the cooler parts of the continent, production of juveniles, usually one summer old, remains the best alternative without cheap heated water, though future studies on the physiology of *A. astacus* and on aquaculture technology may bring new solutions to this problem. Of course, profitability of the culture of crayfishes is also dependent upon the future course of the market price.

In the British Isles, pond stocking and culture of *P. leniusculus* has expanded rapidly in the past decade (Richards 1983, Marren 1986, Lowery and Holdich 1988). The signal crayfish is grown to marketable size using trout ponds or tanks, and are fed a variety of feeds including trout pellets, macrophytic vegetation, and animal materials (Holdich et al. 1990). The growth of stocked signal crayfish in the wild in the British Isles has been described by Hogger (1986).

Figure 63. Production methods of *Pacifastacus leniusculus,* Method 2. Finnish crayfish culture method for raising *Pacifastacus leniusculus* by using heated water and incubation trays according to Järvenpää (1987). After Ackefors (1989).

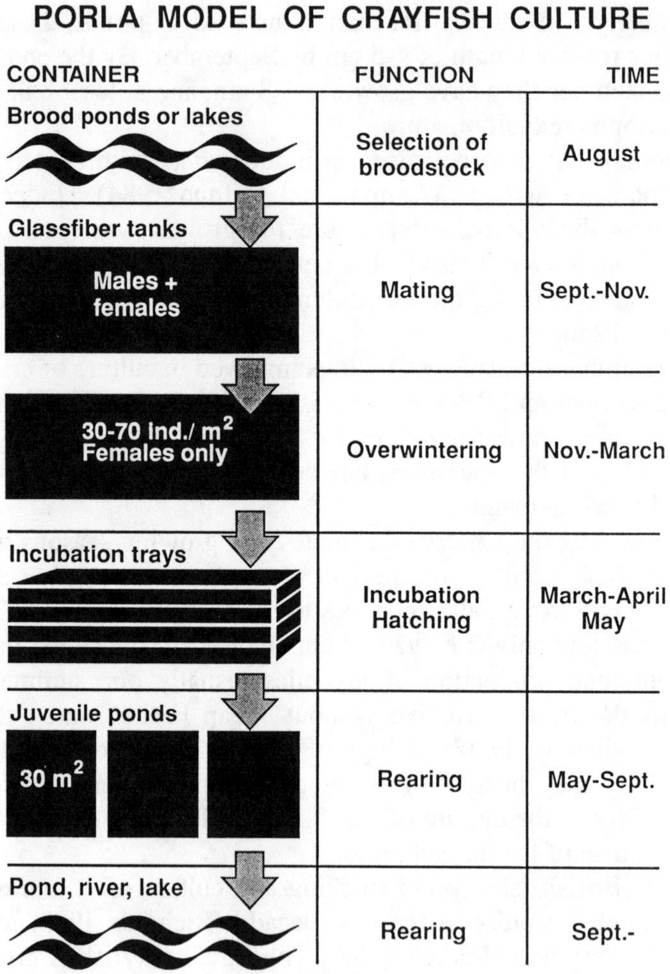

PORLA MODEL OF CRAYFISH CULTURE

CONTAINER	FUNCTION	TIME
Brood ponds or lakes	Selection of broodstock	August
Glassfiber tanks — Males + females	Mating	Sept.-Nov.
30-70 Ind./ m^2 Females only	Overwintering	Nov.-March
Incubation trays	Incubation Hatching	March-April May
Juvenile ponds — 30 m^2	Rearing	May-Sept.
Pond, river, lake	Rearing	Sept.-

CULTURE OF ASTACUS LEPTODACTYLUS

Astacus leptodactylus is the most fecund of all native European species; the number of eggs produced by a female is some 200-400 (Hofmann 1971). The southeasterly distribution indicates that this species tolerates relatively higher water temperatures. The time required for embryonic development there is only 150-200 days. This species seems to tolerate a rather wide range of environmental conditions.

In principle, the culture methods of *A. leptodactylus* do not deviate from those practiced with *A. astacus* and *P. leniusculus*. The culture may be extensive or intensive. The incubation and production of young can be performed in outdoor ponds, keeping the females in troughs or having the stripped eggs in incubation flasks indoors. The best time for stripping is the eyed stage. Such methods for *A. leptodactylus* in Turkey have been described by Köksal (1988). For hatching, 16-18°C seems to be the best temperature range.

Tcherkashina (1977) described her experiments on rearing *A. leptodactylus* in the Rostov area in ponds of 1300 square meter surface area and 1.5 m in depth. The crayfish attained commercial size (10 cm TL) in two summers. In the natural habitat, it takes three to four summers to reach the same size. The ponds were both fertilized and limed, but otherwise the crayfish consumed only natural foods available to them, both planktonic and benthic animals as well as vegetation.

Arrignon (1990) has described several crayfish rearing stations in France, both extensive and intensive, for both *A. leptodactylus* and *A. astacus*. In extensive farms, crayfish are kept in long, narrow earthen outdoor ponds designed to provide maximum shore line. These are at least 30 cm in depth. Sometimes the ponds are lined with metal strips to prevent animals from escaping. The crayfish are usually offered no additional food but they rely on the natural vegetation and animal matter present in the ponds. The ponds may also be constructed of concrete or other materials with the following dimensions: 10 m long, 2.50 m wide, and 0.80 m deep. The edges of the concrete basins or raceways may also be provided with an umbrella-like structure under which the crayfish seek shelter.

In intensive farming, crayfish rearing is done in indoor tanks

188   FRESHWATER CRAYFISH AQUACULTURE

with artificial (tile, etc.) shelters. Arrignon (1990) suggested that
incubation of the stripped eggs in glass jars produces higher hatching success than allowing the females to incubate the eggs themselves. Cukerzis (1979) also obtained better success with stripped
eggs of *A. astacus.*

The main purpose of most French crayfish farming has been
production of young for stocking natural waters. Crayfish culture in
France has regressed recently because the crayfish farms have not
been economically successful. In addition, new government regulations intend to protect native stocks (*A. astacus* and *A. pallipes*) by
restricting the culture and stocking of foreign species, though *P.
clarkii* now occurs widely in the south and west of France. There is
little trapping of wild crayfish and it generates only a few tons
annually (Arrignon et al. 1990; Laurent et al. 1991).

Brodsky (1982) gives a summary of mixed fish/crayfish culture in
the Ukraine. The crayfish, mostly *A. leptodactylus*, are raised in
outdoor ponds with several carp species. Under such conditions the
average fish production may be 2000 kg/ha of pond area and 200 kg
of crayfish. The maximum crayfish production may reach 420 kg/ha.

CULTURE OF AUSTROPOTAMOBIUS SPP.

Austropotamobius pallipes and *A. torrentium* are the smallest of
native European crayfish species, and they have rarely been cultured. Attempts to rear *A. pallipes* in France have been unsuccessful
(Arrignon 1979). In Spain, however, a farm devoted to the culture
of this species was established in El Chaparillo (Cuellar and Coll
1979). The reproductive cycle is rather similar to that of other
European species (Rhodes 1981; Rhodes and Holdich 1979).

Special ponds, 4 × 20 m with a water depth of 1.5 m, are used
for holding broodstock. Smaller and shallower basins with a grated
floor are used for the hatching. The warm climate in southern Spain
shortens the embryonic development and molting periods appear
both in the spring and fall. Hatching of eggs begins as early as
April. The juveniles are transferred to basins where they are fed
with plankton, filamentous algae, daphnids, and diatoms. Later,
ground *Tubifex* and trout starter feed are also added. The adults are

fed boiled potatoes, watercress, nettles, carrots, coarse fish, and commercial fodder for crayfish.

CULTURE OF PROCAMBARUS CLARKII

All the native species of Europe including the introduced *Pacifastacus* belong to the small family Astacidae. The introduced *Procambarus clarkii*, however, belong to the larger family Cambaridae and must live in wet areas, but some species, especially *P. clarkii*, can adapt to temporary dry conditions. Such species occupy all kinds of habitats including subterranean situations, wet meadows, seasonally flooded swamps and marshes, and permanent lakes and streams (Huner 1988). All crayfish–mature and immature–burrow if the habitat dries up regardless of the time of the year, but their abilities to persist is species specific.

Procambarus clarkii was first introduced to Spain in 1973 with two shipments from Louisiana (Habsburgo-Lorena 1979). The introduction of *P. clarkii* in Spain has created a flourishing crayfish industry which by 1980 had an overall monetary turnover of US$7.6 million (Habsburgo-Lorena 1983b). The drawback with a species like *P. clarkii* is its burrowing habit. This activity has caused problems in the irrigation system used for rice farming. During the 1940s rice cultivation started in southern Spain. A shortage of fresh waters in southern Spain has made it necessary to control the distribution of this crayfish as it destroys levees causing water to be drained from the rice fields. Even organophosphate pesticides have been used to prevent an uncontrolled spread. However, construction of sand cores in the levees surrounding the rice paddies seems to be an efficient technique to avoid some of the problems (Habsburgo-Lorena 1983a).

The *P. clarkii* crayfish industry has revitalized the local economy in certain districts of Spain. Several hundred fishermen make their living year round from this fishery, which actually might be considered an extensive form of aquaculture. The main trapping seasons are April and May, and September and October. An antagonism between farmers and fishermen exists, because fishermen are allowed to fish in the farmers' rice fields.

Feasibility of crayfish farming with regard to local ecological considerations has been described by Gaudé (1986). Soil salinity is the major limiting factor for expansion in certain districts. However, aquaculture proper of *P. clarkii* is still not practiced in Spain.

In Spain, 1984 was the fifth consecutive year of drought. Only 4000 of the existing 30,000 hectares of paddy fields were flooded, thereby devastating the crayfish stocks. From an estimated harvest of 3000 tons in 1982, production was reduced to 700 tons in 1983 (Habsburgo-Lorena 1986). The estimated harvest in 1988 was, however, 5000 tons as the drought conditions no longer existed. In normal years, the production of crayfish in rice fields is estimated at 350 kg/ha.

HARVESTING CRAYFISH

In Europe crayfish are harvested almost exclusively with various types of traps. Westman et al. (1979) and Westman et al. (1992) describe typical traps used in Scandinavia and elsewhere in northwestern Europe. These are basket-like in appearance and about 0.5 m long by 0.25 m in diameter. Nylon netting with 1.9 cm square mesh is stretched over connected, concentric rings of wire. The entire trap can be compressed vertically for storage and transportation (Figure 64 and 65). Crayfish enter the traps through funnel-shaped entrances on either end. They are attracted by cut fish, the roach being the most commonly used at least in Finland. Similarly shaped but noncompressible traps made of plastic are gaining popularity now (Figure 66).

Kossakowski (1966) described more traditional European crayfish traps and pots. These were constructed from wicker materials and baited with cut fish.

Spanish crayfish catchers have modified eel traps to catch crayfish without bait. These are elongated tubes about 2 m long and 0.5 m in diameter. There are several funnels that lead to a holding section at the far end from the entrance. The traps are set with the entrances facing downstream. Crayfish move upstream against the current and move into the traps. A panel at the entrance "guides" the crayfish into the trap. For more details about this kind of trap, see Gaudé (1986).

ENVIRONMENTAL FACTORS
AFFECTING CRAYFISH CULTURE

Good water quality is essential for proper growth. Rognerud et al. (1989) listed optimum and tolerance limits for water quality in crayfish culture. Alkalinity should never be lower than 0.1 mEq expressed as mg $CaCO_3$. The calcium concentration needed should exceed 30 mg Ca/l in culture (Fürst 1986).

The role of temperature is, of course, pronounced in crayfish culture, especially in higher latitudes. Cukerzis (1979) summarized the work from his laboratory. Berried females were held for periods ranging from 15 to 60 days at normal winter temperatures of 2-3°C. The temperature was then raised to 8-9°C. After some acclimation time, the females with their attached eggs were kept at a steady temperature of 19-20°C, the normal summertime temperature for Lithuania (cf. Fig. 1). Young typically hatch in June-July, but this thermal regime advanced hatching time 4-5 months to February and March. Such early young reach the size of 120-900 mg by the end of June and make very good stocking material for natural waters where temperatures have already reached 20°C.

According to Cukerzis (1979), a minimum 15-day cold treatment is required in early winter for proper development and survival of the eggs. Thus, the period from egg laying to hatching can be shortened from 7-8 months to 3-4 months. Rhodes (1981) also used elevated temperatures to speed up the development of the eggs of *A. pallipes*.

Reproduction in *A. astacus* starts in October as day length and temperature are decreasing. Molting takes place during the summer when the temperatures are highest (Huner and Lindqvist 1985b; Westin and Gydemo 1986). It ceases when temperature has dropped to 10°C (Ackefors et al. 1989). Westin and Gydemo (1986) report that the trigger mechanism for molting is an increase in temperature and for reproduction it is a decrease in temperature. They suggested that the light regime plays a subordinate role in controlling the two

Figure 64. Typical net mesh Scandinavian crayfish trap–open. J. Huner.

processes. In addition, cycling of the annual thermal cycle twice during the year induces two periods of molting and reproduction. This, then, permits more efficient out-of-season production of young crayfish.

Huner and Lindqvist (1985b) found that reproductively active *A. astacus* captured in August could be held for 2-3 months at 4°C and would still spawn. They noted that shortening of daylength seemed necessary to induce spawning activity in these crayfish, but it was clear that a decrease in temperature was necessary for ovarian de-

Figure 65. Typical net mesh Scandinavian crayfish trap–closed. J. Huner.

velopment to proceed to the full. Their experimental protocol, how-ever, did not permit the fine discrimination between the effects of photoperiod and temperature.

Lahti and Ikäheimo (1979) raised *A. astacus* young in meshed containers in a Finnish lake. They noted that the production of hatchlings per female was very temperature dependent. In a spring with rapidly rising temperatures, the number of hatchlings per fe-male was 92 while in another cold spring the number was only 28. The temperature towards the end of the incubation period is much

Figure 66. Plastic Scandinavian crayfish traps. The two halves are joined at the center. These can be separated and stacked, one half inside the other. J. Huner.

more crucial than at the beginning. For *A. astacus,* Cukerzis (1973) established the optimal incubation temperature at 17-20°C.

Lahti and Ikäheimo (1979) observed that in June the critical thermal maximum for berried females was about 21°C and that death occurred between 23-24°C even though oxygen levels were in excess of 8 ppm. Juvenile *A. astacus* are also sensitive to low temperatures. They fare poorly at 4-5°C. Growth and molting are delayed and there is considerable mortality; however, their resistance to cold temperatures increases by the age of 3 months when low autumn temperatures set in (Lahti and Ikäheimo 1979). Normal intermolt *A. astacus* are certainly more resistant to elevated temperatures, even close to 30°C (Bowler 1963), but at higher latitudes, the condition of wild crayfish due to the highly variable summer temperatures and long winter under the ice may vary highly from year to year and season to season (cf. Lindqvist and Lahti 1983).

Arrignon (1990) indicates that the minimum oxygen level in aquaculture for *A. astacus* would be 5 ppm, but as low as 1 ppm for *A. leptodactylus* and *O. limosus*. Massabuau and Burtin (1984) measured the resting metabolism of *A. leptodactylus* and noted that (at +13°C at least) the oxygen consumption remains the same down to an ambient oxygen concentration of about 1 ppm. Also *A. astacus* may possess a good ability to extract oxygen from very low ambient levels, maybe comparable to that of *A. leptodactylus*, as it often survives the wintertime in apparently very oxygen-poor environments in lake bottom. Yet exact measurements in the laboratory are lacking.

The incipient lethal temperature for *P. leniusculus* is 33°C (Rutledge and Pritchard 1981), and the curve showing the active and standard metabolism (oxygen consumption) at various temperatures shows a decline above 20°C. That is, the scope for activity (= difference between active and standard metabolism) drops off sharply at temperatures above 20°C. This may indicate that the optimum temperature for culture of *P. leniusculus* cannot be much above 20°C. Lozan (1977) found that for *A. astacus, A. leptodactylus, O. limosus*, and *P. leniusculus*, the temperature for maximum locomotor activity and food consumption (per unit body weight) was 20°C in all four species. Comparable data on the energetics of other European crayfish species are mostly lacking, but such knowledge would contribute greatly to the development of better and more profitable culture methods, especially if heated water becomes cheaper and more readily available.

Among the behavioral features of crayfish in aquaculture, cannibalism is one that probably causes the most problems and failures, and contributes to the overall costs of running the hatchery (cf. Mason 1979). The true cause of cannibalistic behavior in crayfish is still awaiting systematic studies, but one could speculate that it is aimed at correcting some imbalance in the available diet (cf. Hird et al. 1986); for instance, crayfish seem to possess a high demand for arginine because their immediate energy source for muscle work is arginine phosphate (Head and Baldwin 1986). In this case cannibalism should be alleviated by providing a more balanced diet. Also, in evolutionary perspective, it may be a hereditary trait with the purpose of lessening competition from other individuals, and in this sense, it may be related to similar phenomena observed in fish (cf. Dionne 1985).

The importance and role of the chemical communication between the brooding female and her offspring, has been shown for some North American crayfish species (Little 1975), but remains to be assessed under culture conditions.

Acidification of natural waters may also set a certain limit to the culture of crayfish as well as to natural populations. Significant inter- and intraspecific differences in exoskeleton mineralization in crayfishes are known to occur; with *A. astacus* this may also reflect the ambient calcium levels (Huner and Lindqvist 1985a). Low pH values affect the reproductive cycle of *A. astacus* as attached eggs may be lost, and there is increased mortality at hatching and during post-larval stages (Appelberg 1984). Low pH also disturbs the mineral balance in the hemolymph of both *A. astacus* and *P. leniusculus* (Appelberg 1985) which may negatively affect survival. Under low ambient pH, the hemolymph pH in *A. astacus* also decreases, but under hypoxia it increases (Nikinmaa et al. 1983). The latter phenomenon may be adaptive in that an increase in pH may increase the oxygen affinity of the hemocyanin.

There have been no systematic studies in Europe on the relationship between acidic waters and crayfish distribution, but crayfish populations have been severely affected in a number of places in

Scandinavia and elsewhere. In Poland, crayfish culture has come to a virtual halt, and many natural populations are suffering from water pollution and acidification (cf. Kossakowski 1973; Kossakowski 1986, personal communication.).

Westman (1985) provides a review of the impact of habitat modification on crayfish fisheries in Europe. This, in many ways, is similar to the impacts on crayfish described in North America (Hobbs and Hall 1974). Lowery and Hogger (1986) presented a detailed study of a specific crayfish population that was destroyed by river works in Great Britain.

CRAYFISH FEED AND NUTRITION

Formulated feeds are generally not used in European crayfish culture, but items such as fish meat, vegetable matter, trout pellets, etc., may be used in addition to what the crayfish obtain from their ponds (plankton, benthic animals, algae, plants, etc.). Our knowledge of what constitutes a balanced diet for any crayfish species is still meager in many respects. This lack of information is also an obstacle to the development of intensive crayfish culture. In North America various types of pellets have been developed and tested for warmwater crayfish (mostly *Procambarus clarkii*) (e.g., Tarshis 1978; Huner and Meyers 1979; Cange et al. 1982; Huner and Barr 1991). D'Abramo et al. (1985) fed an extruded diet with 40% total protein to *P. leniusculus* with good results.

One of the serious problems in trying to compare the results of dietary studies of various species of crustaceans is that there is very little standardization in experimental design, diet formulation, and production techniques, types and purity of feed ingredients for rearing conditions. This has been elucidated in a paper by Castell et al. (1993).

The protein requirement in any feed is a major cost factor. The tests by Huner and Meyers (1979) indicate that at least in *P. clarkii* the gross protein requirement is in the 20-30% range, out of which more than half should be of animal protein substrate in closed systems. Tarshis (1978) obtained higher growth rates with increasing protein up to 50% level in *Procambarus acutus*, another warmwater species, also in closed systems. Huner and Lindqvist (1984) fed rations of 30 and 40% protein to the coldwater species *A. astacus* in tanks and found increased levels of glycogen and lipids in the hepatopancreas of crayfish fed the higher protein feed. Similarly the feed with the higher protein level produced higher concentrations of hemolymph protein which, by analogy to the physiology of lobsters (Castell and Budson 1974), may give rise to better physiological condition and better growth rates.

Standard reference diets with 40%, 31%, and 22% protein content were administered to juvenile *A. astacus* during 14 months in a closed system. The molting frequency, growth rate, and survival rate were highest with 40% protein. They were good with 31% protein if the fat content in the feed was low. A 22% protein level was not enough for good growth and survival (Ackefors et al. 1992). The protein: energy ratios were in the range 114-153 mg protein per kcal. Optimum lipid:carbohydrate ratios changed with protein content in the feed. With increasing physiological age, the animals responded favorably to higher lipid and carbohydrate levels.

Astacus astacus seems to be able to utilize its food even at a low temperature of 2°C, suggesting that winter is also part of its "growing season." The effect was measured by changes in the moisture content of the hepatopancreas (Henttonen and Lindqvist, in preparation). The extent that wintertime feeding adds to the growth increment at molting in summer is in the process of evaluation. Huner et al. (1985) show that the nutritional status of *A. astacus* declines continuously but very slowly at such low temperatures.

Crayfishes are not able to synthesize their own cholesterol (Zandee 1962, 1966) which must be obtained from their food. For *P. leniusculus* the requirement for the sterol component of the feed is at least 0.4% (dry weight) for survival through the juvenile stages (D'Abramo et al. 1985). A mixture of phytosterols was as effective as an equal amount of cholesterol in the partial satisfaction of the requirement; however, phytosterols could not completely replace the cholesterol requirement. Disorders related to sterol metabolism may be a major reason for failures of ecdysis, poor growth, etc., in crayfish culture.

It is known that temperature can affect the fatty acid composition in crustaceans. Farkas and Herodek (1964) showed that the proportions of C_{20}-C_{22} polyunsaturated fatty acids increased in planktonic crustaceans with decreasing temperature, and if fed to fish, the crayfish composition also shifted towards polyunsaturation even at higher temperatures. Acclimation of *A. pallipes* at 4°C caused an increased fatty acid unsaturation of the total phospholipid fraction in the muscle (Cossins 1986). The effect was stronger in an 8-hr light photoperiod than in an 18-hr light photoperiod.

The role of carotenoids in crayfish nutrition is still obscure (cf. Huner and Lindqvist 1984). The lack of such pigments tends to produce blue-colored crayfish in culture (Huner and Meyers 1979).

The vitamin requirements of crustaceans are poorly known (Walker 1975; Conklin 1980) and represent a real challenge to future research which will require new purified diets for testing. D'Abramo and Robinson (1989) summarized the state of knowledge of this field for crayfishes. Even less is known of the requirements for trace elements. It may be mentioned here that Cladocera need Selenium for the synthesis of their cuticle (Keating and Dagbusan 1984), but the requirement of this element by other crustaceans including crayfish remains to be seen.

Goddard (1988) and D'Abramo and Robinson (1989) have written extensive summaries of crayfish nutrition. Other recent reviews on the bioenergetics and nutrition of crustaceans, particularly lobsters, are given by Sasaki et al. (1986), and Castell and Kean (1986). These carry implications for the understanding of crayfish nutrition.

ON CRAYFISH DISEASES

The crayfish "plague," caused by a fungus *Aphanomyces astaci* Schikora, first appeared in Italy in 1860, and it spread rapidly northward with the crayfish trade (Figure 67). In the 1870s it reached both France and Germany, and it destroyed Russian stocks in the 1890s. In 1893, the plague arrived in Finland, and spread further to Sweden in 1907 (cf. Abrahamsson 1969; Westman 1973a). The British Isles were spared until the 1980s (Alderman et al. 1984), and in the early 1980s the plague spread also into Greece. Later in 1984 it was observed in Turkey (Fürst and Söderhäll 1987) and in 1987 it

Figure 67. The spread of the crayfish fungus plague, *Aphanomyces astaci,* in Europe. After Ackefors (1989).

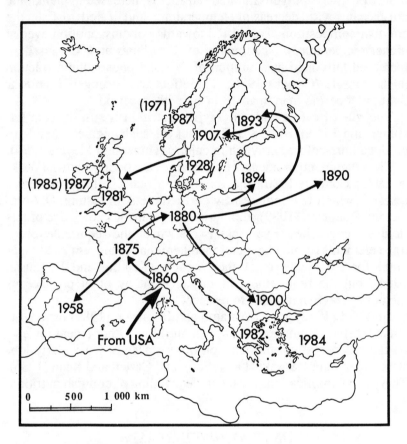

was diagnosed in Ireland (Reynolds 1988). (The true cause of the plague was determined only in the 1930s [Nybelin 1936]; see Smolian [1926] for earlier speculations and studies on the plague.) Apparently the fungus originated in North America, where the crayfish species are relatively resistant to it. Cerenius et al. (1988) provide information about the diagnosis and isolation of this fungus in crayfish and Söderhäll et al. (1988) give a survey of the internal defense mechanisms against disease agents in crayfish.

The life cycle of *A. astaci* exhibits several adaptations to its life as a specialized crayfish parasite. The infective unit of the parasite is the zoospore which is released from the fungal mycelium in the cuticle. These zoospores move by means of their flagella and are able to localize a new crayfish to infect. At close distance to the animal, the orientation of the zoospore is probably aided by the ability of these zoospores to swim towards higher concentrations of the substances released by potential food sources (Cerenius and Söderhäll 1984a). After reaching the crayfish cuticle the zoospores will encyst, i.e., they drop their flagella, round up and become surrounded by highly sticky substances which enable the spore cyst to adhere firmly to its host. If an *A. astaci* zoospore is encysted at other places than the host cuticle they are able to reconvert into the zoospore form by the use of stored material for new flagella, etc. (Cerenius and Söderhäll 1984b, 1985).

These transitions between a swimming zoospore stage and a sessile cyst stage can only be repeated a few times because they require the use of stored material within the spore cell. This storage material cannot be replenished (Cerenius and Söderhäll 1984b). Therefore, the spore stage, including the zoospore-cyst-zoospore transitions, is very shortlived and, only a few days after their formation, the spores which have not found a new crayfish die (Söderhäll 1988). Furthermore, the crayfish plague fungus lacks resting (sexual) spore stages or alternate hosts (Söderhäll 1988). Consequently, if all crayfish in a pond or a natural water system have been killed by the plague, the water is no longer infectious and can be safely restocked.

An interesting development in the *A. astaci* situation is the fact that magnesium chloride can prevent production of sporangia at 25 mM (Rantamäki et al. 1992). A concentration of 200 mM of magnesium chloride will protect *A. astacus* from infection by plague zoospores. This generates interesting possibilities for protecting susceptible crayfish from plague in intensive culture.

The adverse effects of the crayfish plague fungus on native European crayfish stocks cannot be understated. It has now reached Turkey where it has reduced the fishery for *Astacus leptodactylus* by 85% or more (see below) and Great Britain where it is attacking native *Austropotamobius pallipes* stocks (see below). However, the negative impacts of habitat modifications cannot be discounted as even plague-resistant species cannot be used to restore crayfish waters if they are otherwise unsuitable for crayfish.

Westman et al. (1990) list the occurrence of crayfish diseases in some European countries. Bacterial diseases do not seem to be common in crayfish (cf. Unestam 1973), but protozoan and fungal diseases are major problems. The principal microsporidian pathogenic to crayfish is *Thelohania contejeani* which causes the porcelain disease, or white-tail disease. It parasitizes several native species (cf. Cossins 1973; Vey and Vago 1973). It seems to be more common in warmer climates than in the north; however, its true significance to crayfish population dynamics is unknown.

The true taxonomic position of another possible "protozoan" parasite, *Psorospermium haeckeli*, remains to be determined (Vey 1979; Nylund et al. 1983). It appears to have some negative impact on native populations of crayfish.

Of potential interest for crayfish farmers are the recent findings (Kobayashi and Söderhäll 1990; Cerenius et al. 1991) that *P. haeckeli* elicits cellular and humoral immune reactions in both *P. leniusculus* and *A. astacus*. The presence of *P. haeckeli* may weaken the ability of the crayfish to mount an efficient defense against other intruding pathogens. This hypothesis is strengthened by the observations in Turkey of *A. leptodactylus* showing that mortalities associated with crayfish plague were higher in populations where both *P. haeckeli* and the plague fungus were present compared to populations with the fungus only (Fürst and Söderhäll 1987).

Among the fungal diseases, the burn spot disease in *A. astacus* is usually said to be caused by *Ramularia astaci*, but it is caused by several species of the genus *Fusarium* (Prof. Kenneth Söderhäll, Uppsala University, Uppsala, Sweden). Similar fungi parasitize other crayfish species (Unestam 1973). Vey (1977) described fungi that infect crayfish eggs under rearing conditions.

The major threat to native European crayfish populations comes from the crayfish plague fungus, *Aphanomyces astaci*. All native European species are susceptible to it (Unestam 1969), but in the North American *P. leniusculus*, it elicits only a defense-like reaction in the cuticle with melanization around the hyphae and occasional loss of a walking leg or pleopod (Unestam 1972); yet under stressful situations in the wild as well as in culture the signal crayfish may succumb to the disease. The main vector of this disease is probably the fishermen and traders who move contaminated gear or infected crayfish from place to place. In places like central Finland with long watercourses, the disease can spread over long distances from population to population (Westman and Nylund 1979).

The crayfish fungus plague may enter a crayfish culture station through *Aphanomyces* spores from a nearby river or lake through flowing water if the site is occupied by infected wild crayfish. Thus, it may be advisable to use recirculated water, to rely on spring water, or to build the facility next to a water source devoid of crayfish. Several countries have imposed measures for controlling the spread of the disease including disinfection of crayfish traps with formalin and bleach and banning imports of live crayfish.

The crayfish plague fungus kills the susceptible crayfish every time an infection has occurred. The time from attachment of the spore and penetration of the hyphae until death of the host is about a week during the summer growing season. It may be longer at lower temperatures (cf. Unestam et al. 1977).

Observations on the course of the disease in *A. astacus* have been made in an aquarium at about 22°C (Lindqvist unpublished.) (See Figure 68.) Soon after the infection has taken place, the crayfish start showing increased locomotor activity, move upwards towards light, and make cleaning movements with their walking legs. When placed upside down on a table, they are able to straighten up in a few seconds until the fifth or sixth day. At that time, they start exhibiting bending movements and have a tendency to tip over. There is often a loss of legs or chelipeds. Paralysis and death follow and the animal is usually found lying on its back. The hyphae grow out of the crayfish body and produce swimming zoospores which

Figure 68. Symptoms of acute crayfish fungus plague in noble crayfish. O. Lindqvist.

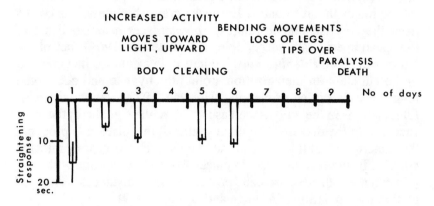

survive for a few days (Unestam 1973). The spores are the source of infection, but the infected crayfish through their increased locomotor activity can also spread the disease more effectively to other crayfish.

The commensal helminth (*Branchiobdella* spp.) commonly occur on the outside cuticle of *A. astacus* and *P. leniusculus*. Their true effects on the host, if any, are not known.

FUTURE PROSPECTS

The future of crayfish culture in Europe is certainly variable in different parts of the continent, but for the near future one can expect a steady expansion. After the collapse of the Turkish fishery, its place may be ultimately taken by Spain where *P. clarkii* is harvested from feral populations and little real culture is practiced. Spanish crayfish production seems to be on a relatively favorable socioeconomic footing (Habsburgo-Lorena 1983b). At least for the time being, imports from Louisiana seem to be on the increase to meet the demand in Europe, though the large-scale acceptance of *P. clarkii* by the European consumer is still an open question.

Intensive and semi-intensive crayfish culture still need more development of production techniques before they become truly profitable. The production of juveniles, especially in cooler climates, appears to be profitable where the fisheries authority has taken an active role in restoring crayfish waters. In some countries the main aim seems to be the restoration of the native stocks. Carrying out the whole life cycle in commercial aquaculture still faces many economic problems, but with future research they should be overcome. In Scandinavia where there are still plenty of suitable rivers and lakes for stockings, the market for crayfish juveniles is expanding. For instance, in Finland, the production of juveniles in the next few years may be several millions per year if all the private facilities now planned are completed. Sweden is the leading country in Scandinavia for the number of existing crayfish culture facilities.

In the past the European research effort in the field of crayfish culture has been directed more toward development and protection of natural populations. There has been little scientific effort toward active aquaculture especially when compared to the situation in North America. Yet many of the problems facing crayfish culture are common to both Europe and North America. More information is needed on population biology, nutrition, energetics, harvesting technology, diseases, etc. Interest in crayfish culture is increasing in the private sector in Europe. It is expected to expand partly as a result of the collapse of the Turkish fishery. The strength of the economics of the efforts will depend on the development of the markets as well as on better and less risky production methods, including the use of heated water. So far, nearly all attempts at crayfish culture in Europe have been of small scale. As a result of improving culture technology and expanding markets, we may expect that large-scale businesses with the accompanied benefits of scale will gradually develop.

LITERATURE CITED

Abrahamsson, S. 1969. Historik, allman oversikt och reformforslag betraffande fangst och handel med flodkrafta. *Fauna och Flora* 64:98-104.

Abrahamsson, S. 1971. Density, growth and reproduction in populations of *Astacus astacus* and *Pacifastacus leniusculus* in an isolated pond. *Oikos* 22: 373-380.

Abrahamsson, S. 1972. Fecundity and growth of some populations of *Astacus*

astacus Linne in Sweden, with special regard to introduction in the northern Sweden. *Institute of Freshwater Research, Drottningholm*, Report No. 52: 23-37.

Abrahamsson, S. 1973a. The crayfish *Astacus astacus* in Sweden and the introduction of the American crayfish *Pacifastacus leniusculus*. *Freshwater Crayfish* 1:27-40.

Abrahamsson, S. 1973b. Methods for restoration of crayfish waters in Europe. *Freshwater Crayfish* 1:203-210.

Abrahamsson, S., and C. R. Goldman. 1970. The distribution, density and production of the crayfish *Pacifastacus leniusculus* (Dana) in Lake Tahoe, California-Nevada. *Oikos* 21:83-91.

Ackefors, H. 1989. Intensification of European freshwater crayfish culture in Europe. *Special session on crayfish culture of Aquaculture 1989*, World Aquaculture Society, Los Angeles, USA, February 13, 1989, 29 pp (mimeo.).

Ackefors, H. 1991. Kräftfiske och kräftodling i Sverige (Crayfish fishing and crayfish cultivation in Sweden). Lecture given at V. Fiskerimuseiöreningens publikationer 5, 23-29.

Ackefors, H., J. D. Castell, L. D. Boston, P. Räty, and M. Svensson. 1992a. Standard experimental diets for crustacean nutrition research. II. Growth and survival of juvenile crayfish *Astacus astacus* (Linne) fed diets containing various amounts of protein, carbohydrate and lipid. *Aquaculture* 104:341-356.

Ackefors, H., R. Gydemo, and L. Westin. 1989. Growth and survival of juvenile crayfish, *Astacus astacus*, in relation to food and density. Pages 365-373 in, N. De Pauw, E. Jaspers, H. Ackefors, and N. Wilkins (eds.). *Biotechnology in progress*. European Aquaculture Society, Bredene, Belgium.

Alderman, D. J., J. L. Polglase, M. Frayling, and J. Hogger. 1984. Crayfish plague in Britain. *Journal of Fish Diseases* 7:401-405.

Appelberg, M. 1984. Early development of the crayfish *Astacus astacus* L. in acid water. *Reports of the Institute of Freshwater Research*, Drottningholm 61: 48-59.

Appelberg, M. 1985. Changes in haemolymph ion concentration of *Astacus astacus* L. and *Pacifastacus leniusculus* (Dana) after exposure to low pH and aluminum. *Hydrobiologia* 121:19-25.

Arrignon, J. 1979. *L'écrevisse et son élevage*. Gauthier-Villars. Paris. (Collection "Nature et Agriculture"), France.

Arrignon, J. 1990 (2nd ed.). *L'écrevisse et son élevage*. Gauthier-Villars. Paris. (Collection "Nature et Agriculture"), France.

Arrignon, J. V., J. V. Huner, and P. J. Laurent. 1990. *L'Écrevisse rouge des marais*. Le Technicien D'Agriculture Tropicale, Editions Maison neuve et LaRose, Paris, France.

Auvergne, A. 1982. *El Cangrejo de Rio*. Cria y Explotacion. Ediciones Mundi-Preusa, Barcelona, Spain.

Bowler, K. 1963. A study of the factors involved in acclimatization to temperature and death at high temperatures in *Astacus pallipes*. I. Experiments on intact animals. *Journal of Cellular and Comparative Physiology* 62:119-132.

Brinck, P. 1977. Developing crayfish populations. *Freshwater Crayfish* 3:211-228.

Brinck, P. 1983. Sture Abrahamsson Memorial Lecture. *Freshwater Crayfish* 5:xxi-xxxvii.

Brodsky, S. Y. 1982. Sur l'élevage en Union Sovietique des ecrevisses de riviere (Astacidae) par la methode industrielle et sur les perspectives de cette methode. *La Pisciculture Francaise* 18(78):5-9.

Brodsky, S. Y. 1983. On the systematics of Palearctic crayfishes (Crustacea, Astacidae). *Freshwater Crayfish* 5:464-470.

Cange, S., M. Miltner, and J. W. Avault, Jr. 1982. Range pellets as supplemental crayfish feed. *Progressive Fish-Culturist* 44(1):23-24.

Castell, J. D., and S. D. Budson. 1974. Lobster nutrition: The effect on *Homarus americanus* of dietary protein levels. *Journal of the Fisheries Research Board of Canada* 31:1363-1370.

Castell, J. D., and J. C. Kean. 1986. Evaluation of the role of nutrition in lobster recruitment. *Canadian Journal of Fisheries and Aquatic Sciences* 43:2320-2327.

Castell, J. D., H. Ackefors, L. D. Boston, D. J. Scarratt, and X. Xu. 1993. Standard experimental diets for crustaceans nutrition research. I. Experimental design and research techniques. (manuscript).

Cedrins, R. 1985. Odling av sötvattenskraftor i Sverige. *Nordisk Aquakultur* 1(5):14-17.

Cerenius, L., and K. Söderhäll. 1984a. Chemotaxis in *Apahnomyces astaci* an arthropod parasitic fungus. *Journal of Invertebrate Pathology* 42:278-281.

Cerenius, L., and K. Söderhäll. 1984b. Repeated zoospore emergence from isolated spore cysts of *Aphanomyces astaci*. *Experimental Mycology* 8:370-377.

Cerenius, L., and K. Söderhäll. 1985. Repeated zoospore emergence as a possible adaptation to parasitism in *Aphanomyces*. *Experimental Mycology* 9:259-263.

Cerenius, L., P. Henttonen, O. V. Lindqvist, and K. Söderhäll. 1991. The crayfish pathogen *Psorospermium haeckili* activates the prophenoloxidase activating system of freshwater crayfish in vitro. *Aquaculture* 99:225-233.

Cerenius, L., K. Söderhäll, M. Persson, and R. Ajaxon. 1988. The crayfish plague fungus *Aphanomyces astaci*–diagnosis, isolation, and pathobiology. *Freshwater Crayfish* 7:131-144.

Conklin, D. E. 1980. Nutrition. Pages 277-300 in J. S. Cobb and A. F. Phillips (eds.). *The biology and management of lobsters*. I. Academic Press, New York.

Cossins, A. R. 1973. *Thelohania contejeani* Henneguy, microsporidian parasite of *Austropotamobius pallipes* Lereboullet–an histological and ultrastructural study. *Freshwater Crayfish* 1:151-164.

Cossins, A. R. 1986. Changes in muscle lipid composition and resistance adaptation to temperature in the freshwater crayfish, *Austropotamobius pallipes*. *Lipids* 11(4):307-316.

Cuellar, L., and M. Coll. 1979. First essays of controlled breeding of *Astacus pallipes* (Ler.). *Freshwater Crayfish* 4:273-286.

208   *FRESHWATER CRAYFISH AQUACULTURE*

Cuellar, L., and M. Coll. 1983. Epizootiology of the crayfish plague (aphanomycosis) in Spain. *Freshwater Crayfish* 5:545-548.

Cukerzis, J. 1959. Zählmethoden zur Bestimmung, sowie Aufzuchtverfahren und Schonmassnahmen zur Hebung der Edelkrebsbestände Litauens. *Ucenye Zapiski Vil'njuskogo Ucitel'skogo Instituta* 1:143-163. (Übersetzung Nr. 140, Statens Naturvetenskapliga Forskningsråd.)

Cukerzis, J. 1973. Biologische Grundlagen der Methode der künstlichen Aufzucht der Brut des *Astacus astacus* L. *Freshwater Crayfish* 1:187-201.

Cukerzis, J. 1979 (ed.). *Biology of the crayfish of the Lithuanian inner waters.* Vilnius. (In Russian with English summaries).

Cukerzis, J. 1984. *La biologie de l'écrevisse* (*Astacus astacus* L.). I.N.R.A., Paris, France.

Cukerzis, J. M. 1988. *Astacus astacus* in Europe. Pages 309-340 in: D. M. Holdich and R. S. Lowery (eds.). *Freshwater crayfish–biology, management and exploitation.* Croom Helm Ltd.

D'Abramo, L., J. S. Wright, K. H. Wright, C. E. Bordner, and D. E. Conklin. 1985. Sterol requirements of cultured juvenile crayfish, *Pacifastacus leniusculus.* *Aquaculture* 49:245-255.

D'Abramo, L., and E. H. Robinson. 1989. Nutrition of crayfish. CRC Critical Reviews in *Aquatic Sciences* 1(4):711-728.

de Bikuna, B. G., L. Docampo, and R. Asensio. 1989. Distribution et autoecologie de l'écrevisse a pattes blanches, *Austropotamobius pallipes* (Ler.) a Bizkaia (Pays Basque, Espagne). *Annales de Limnologie* 25(3):219-229.

Dionne, M. 1985. Cannibalism, food availability, and reproduction in the mosquito fish (*Gambusia affinis*): A laboratory experiment. *American Naturalist* 126:16-23.

Erencin, Z., and G. Köksal. 1977. On the crayfish, *Astacus leptodactylus*, in Anatolia. *Freshwater Crayfish* 3:187-192.

Farkas, T., and S. Herodek. 1964. The effect of environmental temperature on the fatty acid composition of crustacean plankton. *Journal of Lipid Research* 5:369-373.

Fürst, M. 1977. Introduction of *Pacifastacus leniusculus* (Dana) into Sweden: methods, results and management. *Freshwater Crayfish* 3:229-247.

Fürst, M. 1986. Kräftodling i dammar. *Information från Sötvattenslaboratoriet Drottningholm.* 3:1-34.

Fürst, M. 1990. National report on Sweden. In K. Westman, M. Pursiainen and P. Westman (eds.). 1990. Status of crayfish stocks, fisheries, diseases and culture in Europe. *Report of the FAO European Inland Fisheries Advisory Commission (EIFAC) Working Party on Crayfish.* Finnish Game and Fisheries Institute Report No. 3, 1990, Helsinki, Finland, Yliopistopaino, Helsinki.

Fürst, M., and K. Söderhäll. 1987. Report of the crayfish *Astacus leptodactylus* in Turkey. Diseases and present distribution of the crayfish plague *Aphanomyces astaci.* FAO unpublished report.

Gaudé, A. P., III. 1986. Ecology and production of Louisiana red swamp crayfish *Procambarus clarkii* in southern Spain. *Freshwater Crayfish* 6:111-130.

Goddard, J. S. 1988. Food and feeding. In D. M. Holdich and R. S. Lowery (eds.). pp. 145-146. *Freshwater Crayfish Biology, Management and Exploitation.* Croom Helm, London.

Goddard, J. S., and J. B. Hogger. 1986. The current status and distribution of freshwater crayfish in Britain. *Field Studies* 6:383-396.

Groves, R. E. 1985. *The crayfish. Its nature and nurture.* Fishing News Books Ltd., Farnham, Surrey, England, 72 pp.

Gydemo, R., and L. Westin. 1989. Growth and survival of juvenile *Astacus astacus* L. at optimized water conditions. Pages 383-392 in N. De Pauw, E. Jaspers, H. Ackefors, and N. Wilkins (eds.). *Biotechnology in progress.* European Aquaculture Society, Bredene, Belgium, 2 volumes, 1231 pp.

Habsburgo-Lorena, A. 1979. Crayfish situation in Spain, a note in the *International Association of Astacology Newsletter* 3(2):1-2, July 1979.

Habsburgo-Lorena, A. S. 1983a. Some observations on crawfish farming in Spain. *Freshwater Crayfish* 5:549-551.

Habsburgo-Lorena, A. 1983b. Socioeconomic aspects of the crawfish industry in Spain. *Freshwater Crayfish* 5:552-554.

Habsburgo-Lorena, A. S. 1986. The status of the *Procambarus clarkii* population in Spain. *Freshwater Crayfish* 6:131-133.

Habsburgo-Lorena, A. 1990. National report on Spain. In K. Westman, M. Pursiainen and P. Westman (eds.). 1990. Status of crayfish stocks, fisheries, diseases and culture in Europe. *Report of the FAO European Inland Fisheries Advisory Commission (EIFAC) Working Party on Crayfish.* Finnish Game and Fisheries Institute Report No. 3, 1990, Helsinki, Finland, Yliopistopaino, Helsinki.

Haeckel, E. 1857. Über die Gewebe des Flusskrebses. *Archiven fur Anatomie und Physiologie der wissenschaftlichen Medizine* 1857:469-568.

Hammarlund, C. G., and A. S. Karlsson. 1992. *Litet ABC om kräftor och kräftodling. Fiskeriverket,* Sweden, 22 pp. (5: ereviderode upplagan).

Head, G., and J. Baldwin. 1986. Energy metabolism and the fate of lactate during recovery from exercise in the Australian freshwater crayfish *Cherax destructor. Australian Journal of Marine and Freshwater Research* 37:641-646.

Herfort-Michieli, T. 1973. Der Krebsbestand und seine Erneuerung in Slowenien. *Freshwater Crayfish* 1:97-104.

Hessen, D.O., T. Taugbøl, E. Fjeld, and J. Skurdal. 1987. Egg development and lifecycle timing in the noble crayfish (*Astacus astacus*). *Aquaculture* 64:77-82.

Hird, F. J. R., S. C. Cianciosi, and R. M. McLean. 1986. Investigations on the origin and metabolism of the carbon skeleton of ornithine, arginine and proline in selected animals. *Comparative Biochemistry and Physiology* 83B:179-184.

Hobbs, H. H., Jr., and E. T. Hall, Jr. 1974. Crayfishes (Decapoda: Astacidae). Pages 195-214 in *Pollution ecology of freshwater invertebrates.* Academic Press, New York.

Hofmann, J. 1971 (1980). *Die Flusskrebse. Biologie, Haltung und wirtschaftliche Bedeutung.* Paul Parey, Hamburg, Germany.

Hogger, J. B. 1986. Aspects of the introduction of "signal crayfish," *Pacifastacus leniusculus* (Dana), into the southern United Kingdom. I. Growth and survival. *Aquaculture* 58:27-44.

Holdich, D. M., and R. S. Lowery. 1988. *Freshwater crayfish. Biology, management and exploitation.* Croom Helm, London.

Holdich, D. M., K. Bowler, and R. Lowery. 1990. In K. Westman, M. Pursiainen and P. Westman (eds.). 1990. Status of crayfish stocks, fisheries, diseases and culture in Europe. *Report of the FAO European Inland Fisheries Advisory Commission (EIFAC) Working Party on Crayfish.* Finnish Game and Fisheries Institute Report No. 3, 1990, Helsinki, Finland, Yliopistopaino, Helsinki.

Huner, J. V. 1984. *Freshwater crayfish–a growing supply.* International Shellfish Conference, U.K. (Jersey), 22-24 May 1984, HGB Heighway House, London. Summary.

Huner, J. V. 1988. Status of crayfish transplantations. *Freshwater Crayfish* 7:29-34.

Huner, J. V., and J. E. Barr. 1991. *Red swamp crawfish, biology and exploitation.* Louisiana State University, Baton Rouge, Louisiana 70803, USA.

Huner, J. V., and O. V. Lindqvist. 1984. Effects of temperature and diet on reproductively active male noble crayfish (*Astacus astacus*) subjected to bilateral eyestalk ablation. *Journal of the World Mariculture Society* 15:138-141.

Huner, J. V., and O. V. Lindqvist. 1985a. Exoskeleton mineralization in astacid and cambarid crayfishes (Decapoda, Crustacea). *Comparative Biochemistry and Physiology* 80A:515-521.

Huner, J. V., and O. V. Lindqvist. 1985b. Effects of temperature and photoperiod on mating and spawning activities of wild caught noble crayfish (*Astacus astacus* Linne) (Astacidae, Decapoda). *Proceedings of the World Aquaculture Society* 16:225-226.

Huner, J. V., and O. V. Lindqvist. 1986. A stunted crayfish *Astacus astacus* population in Central Finland. *Freshwater Crayfish* 6:156-165.

Huner, J. V., P. Henttonen, and O. V. Lindqvist. 1991. Length-length and length-weight characterizations of noble crayfish, *Astacus astacus* L. (Decapoda, Astacidae), from central Finland. *Journal of Shellfish Research* 10:195-196.

Huner, J. V., O. V. Lindqvist, and H. Könönen. 1985. Responses of intermolt noble crayfish, *Astacus astacus* (Decapoda, Astacidae), to short-term and long-term holding conditions at low temperature. *Aquaculture* 47:213-221.

Huner, J. V., and S. P. Meyers. 1979. The dietary protein requirements of the red crawfish, *Procambarus clarkii* (Girard) (Decapoda, Cambaridae), grown in a closed system. *Proceedings of the World Aquaculture Society* 10:751-760.

Huxley, T. H. 1880. *The crayfish. An introduction to the study of zoology.* D. Appleton and Co., London, Great Britain.

Jay, D., and D. M. Holdich. 1981. The distribution of the crayfish, *Austropotamobius pallipes*, in British waters. *Freshwater Biology* 11:121-129.

Järvenpää, T. 1987. Signalkräftodling vid Porla Fiskodlingsanstalt i Finland. (Cultivation of signal crayfish at Porla Rearing unit in Finland). Lecture given at Malmö exhibition center, Sweden, February 1987.

Järvenpää, T., and P. Ilmarinen. 1990. Artificial incubation of crayfish eggs on moving trays. Abstract. 8th International Symposium of Astacology. Baton Rouge, Louisiana.

Järvi, T. H. 1910. Ravusta ja rapukulkutaudeista Suomessa. *Suomen Kalastuslehti* 19:73-90.

Karaman, M. S. 1962. Ein Beitrag zur Systematik der Astacidae (Decapoda). *Crustaceana* 3:173-191.

Karlsson, A. S. 1977. The freshwater crayfish. *Fish Farming International* 4(6):3,5.

Keating, K. I., and B. C. Dagbusan. 1984. Effect of selenium deficiency on cuticle integrity in the Cladocera (Crustacea). *Proceedings of the National Academy of Science* (USA) 81:3433-3437.

Kobayashi, M., and K. Söderhäll. 1990. Comparison of concanavalin-A reactive determinants on isolated haemocytes of parasite-infected and non-infected crayfish. *Diseases of Aquatic Organisms* 9:141-147.

Köksal, G. 1988. *Astacus leptodactylus* in Europe. Pages 365-400 in D. M. Holdich and R. S. Lowery (eds.). *Freshwater crayfish–biology, management and exploitation.* Croom Helm Ltd.

Kossakowski, J. 1966. *Crayfish ("Raki"),* Panstowe Wydawnectwo Rolinicze i Lesne, Warzawa, Poland, 292 pp.–Available from U.S. Department of Commerce, Washington, DC as translation TT 70-55114, 163 pp.

Kossakowski, J. 1973. The freshwater crayfish in Poland. *Freshwater Crayfish* 1:17-26.

Kusnetzow, I. D. 1898. Fischerei und Thierbeutung in den Gewassern Russlands. St. Petersburg.

Lagerqvist, L. O., and E. Nathorst-Böös. 1980. *En liten bok om kräftor.* Liber, 86 pp.

Lahti, E., and M. Ikäheimo. 1979. Ravun matimunien haudontakokeista. *Suomen Kalastuslehti* 86(4):80-82.

Laubier, L., and O. V. Lindqvist. 1990. New developments in shrimp and crayfish culture. In N. De Pauw and R. Billard (eds.). *Aquaculture Europe 1989–Business joins science 1990.* European Aquaculture Society Special Publication No. 12, Bredene, Belgium.

Laurent, P. 1973. *Astacus* and *Cambarus* in France. *Freshwater Crayfish* 1:69-78.

Laurent P. J. 1988. *Austropotamobius pallipes* and *A. torrentium,* with observations on their interaction with other species in Europe. In D. M. Holdich and R. S. Lowery (eds.). pp. 341-364. *Freshwater Crayfish Biology, Management and Exploitation.* Croom Helm, London.

Laurent, P. J., and J. Forest. 1979. Donnees sur les écrevisses qu' on peut recontrer en France. *La Pisciculture Francaise* 15(56):25-37.

Laurent, P. J., and M. Suscillon. 1962. Les écrevisses en France. *Annales de la Station Centrale d'Hydrobiologie Appliquee* 9:333-395.

Laurent, P. J., J. Escomel, and P. Laurent. 1979. Premiers resultats des introductions expérimentales en eaux closes de *Pacifastacus leniusculus* Dana. *La Pisciculture Francaise* 15(56):51-57.

Laurent, P. J., H. Leloirn, and A. Neveu. 1991. Remarques sur l'acclimation en

France de *Procambarus clarkii* (Decapoda Cambaridae). *Bull. Mens. Soc. Linn. Lyon.* 60:166-173.

Lindqvist, O. V. 1988. Restoration of native European crayfish stocks. *Freshwater Crayfish* 7:6-12.

Lindqvist, O. V., and E. Lahti. 1983. On the sexual dimorphism and condition index in the crayfish, *Astacus astacus* L. in Finland. *Freshwater Crayfish* 5:3-11.

Little, E. E. 1975. Chemical communication in maternal behaviour of crayfish. *Nature* 255:400-401.

Lowery, R. S., and J. B. Hogger. 1986. The effect of river engineering works and disease on a population of *Austropotamobius pallipes* in the river Lea, UK. *Freshwater Crayfish* 6:94-99.

Lowery, R. S., and D. M. Holdich. 1988. *Pacifastacus leniusculus* in North America and Europe, with details of the distribution of introduced and native crayfish species in Europe. Pages 283-308 in D. M. Holdich and R. S. Lowery (eds.). *Freshwater crayfish management, biology, management, and exploitation.* Croom Helm, London and Sydney, Timber Press, Portland, Oregon, USA.

Lozan, J. L. 1977. Experientelle Untersuchungen zum Bewegungs-verhalten von vier Astaciden *Astacus astacus* (Linné 1758), *Astacus leptodactylus* (Eschholz 1823), *Orconectes limosus* (Rafinésque 1817), *Pacifastacus leniusculus* (Dana 1852). *Diplomarbeit.* Hamburg, June 1977.

Mancini, A. 1986. Astaciculture allevamento e pesca dei gamberi d'acqua dolce. Edagricole, Bologna, Italy, 180 pp.

Marren, P. 1986. The lethal harvest of crayfish plague. *New Scientist* 109(1495): 46-50.

Mason, J. C. 1975. Crayfish production in a small woodland stream. *Freshwater Crayfish* 2:449-479.

Mason, J. C. 1979. Effects of temperature, photoperiod, substrate, and shelter on survival, growth, and biomass accumulation of juvenile *Pacifastacus leniusculus* in culture. *Freshwater Crayfish* 4:73-82.

Massabuau, J. C., and B. Burtin. 1984. Regulation of oxygen consumption in the crayfish *Astacus leptodactylus* at different levels of oxygenation: role of peripheral O_2 chemoreception. *Journal of Comparative Physiology* B 155:43-49.

Morales, J. C. 1987. *Cria del cangrejo de rio.* Editorial Hispano European, S. A., Barcelona, Spain.

Moriarty, C. 1973. A study of *Austropotamobius pallipes* in Ireland. *Freshwater Crayfish* 1:57-67.

Munkhammar, T., R. Gydemo, L. Westin and H. Ackefors. 1989. Survival of noble crayfish (*Astacus astacus* L.) larvae alone and in the presence of females. Pages 409-414 in N. De Pauw, E. Jaspers, H. Ackefors, and N. Wilkins (eds.). *Biotechnology in progress.* European Aquaculture Society, Bredene, Belgium.

Nikinmaa, M., T. Järvenpää, K. Westman, and A. Soivio. 1983. Effects of hypoxia and acidification on the haemolymph pH values and ion concentrations in the freshwater crayfish (*Astacus astacus* L.). *Finnish Fisheries Research* 5:17-22.

Nolte, W. 1933. Über die Möglichkeit, saure Gewässer mit Krebsen zu bewirtschaften. *Fischerei-Zeitung* 36(49).

Nordqvist, O. 1898. Dissenting opinion (Vastalause). Pages 219-221 in *Komiteanmietinto No. 4, Kalastuskomitean mietintö (Committee Report on Fisheries*, in Finnish). Helsinki 1898.

Nybelin, O. 1936. Untersuchungen über die Ursache der in Schweden gegenwartig vorkommenden Krebspest. *Reports of the Institute of Freshwater Research*, Drottningholm (Sweden) 9:3-29.

Nylund, V., K. Westman, and K. Laintmaa. 1983. Ultrastructure and taxonomic position of the crayfish parasite *Psorospermium haeckeli* Hilgendorf. *Freshwater Crayfish* 5:307-314.

Persson, M., and K. Söderhäll. 1983. *Pacifastacus leniusculus* Dana and its resistance to the parasitic fungus *Aphanomyces astaci. Freshwater Crayfish* 5:292-298.

Persson, M., L. Cerenius, and K. Söderhäll. 1987. The influence of haemocyte number on the resistance in the freshwater crayfish *Pacifastacus leniusculus* in the parasitic fungus *Apahnomyces astaci. Journal of Fish Diseases* 10:471-477.

Pursiainen, M., and T. Järvenpää. 1981. Ravun viljely. *Suomen Kalankasvattaja* 10(4):43-45.

Pursiainen, M., T. Järvenpää, and K. Westman. 1983. A comparative study on the production of crayfish (*Astacus astacus*) juveniles in natural food ponds and by feeding in plastic basins. *Freshwater Crayfish* 5:392-402.

Pursiainen, M., and M. Saarela. 1985. Uusia menetelmiä ravunviljelyssä. *Suomen Kalankasvattaja* 14(4):60-61.

Rahe, R., and E. Soylu. 1989. Identification of the pathogenic fungus causing destruction of Turkish crayfish species (*Astacus leptodactylus*). *Journal of Invertebrate Pathology.* 54:10-15.

Ramos, M. A., and T. M. G. Pereira. 1981. Un novo Astacidae para a fauna Portuguesa: *Procambarus clarkii* (Girard, 1852). Biol. Inst. Nacional Invest. *Pescas Lisboa.* 6 Jul.-Aug.:37-41.

Rantamäki, J. L., J. Cerenius, and K. Söderhäll. 1992. Prevention of transmission of the crayfish plague fungus (*Aphanomyces astaci*) to the freshwater crayfish *Astacus astacus* by treatment with MgCl$_2$. *Aquaculture* 104:11-18.

Reynolds, J. D. 1988. Crayfish extinctions and crayfish plague in Central Ireland. *Biological Conservation* 45:279-285.

Rhodes, C. P. 1981. Artificial incubation of the eggs of the crayfish *Austropotamobius pallipes. Aquaculture* 17:345-358.

Rhodes, C. P., and D. M. Holdich. 1979. On size and sexual dimorphism in *Austropotamobius pallipes* (Lereboullet). *Aquaculture* 25:345-358.

Richards, K. J. 1983. The introduction of the signal crayfish into the United Kingdom and its development as a farm crop. *Freshwater Crayfish* 5:557-563.

Rognerud, S., M. Appelberg, A. Eggereide, and M. Pursiainen. 1989. Water quality and effluents. In Skurdal, J., K. Westman, and P. I. Bergan (eds.).

214   FRESHWATER CRAYFISH AQUACULTURE

Crayfish culture in Europe. Report from the workshop on crayfish culture, 16-19 Nov., 1987. Trondheim, Norway.

Rutledge, P. S., and A. W. Pritchard. 1981. Scope for activity in the crayfish *Pacifastacus leniusculus*. *American Journal of Physiology* 240:R87-R92.

Sasaki, G. C., J. M. Capuzzo, and P. Biesiot. 1986. Nutritional and bioenergetic considerations in the development of the American lobster *Homarus americanus*. *Canadian Journal of Fisheries and Aquatic Sciences* 43:2311-2319.

Schweng, E. 1973. *Orconectes limosus* in Deutschland, insbesonders im Rheingebiet. *Freshwater Crayfish* 1:79-87.

Sestokas, J., and J. M. Cukerzis. 1973. The effect of phenol on crayfish. Biol. issled. na vnutr. vod. Pribaltik: Tr. Xl nauch. Conf. Minsk, pp. 120-121.

Skurdal, J., and D. O. Hessen. 1985. Ferskvannkrepsen-vårt nye oppdrettsdyr. *Forskningsnytt* 30(3):5-9.

Smolian, K. 1926. Der Flusskrebs, seine Verwandte und die Krebsgewässer. Pages 423-524 in R. Demoll and H. N. Maier (eds.). *Handbuch der Binnenfischerei Mitteleuropas*, Volume 5:423-524. Stuttgart, Germany.

Söderhäll, K. 1988. Fungal parasites and other diseases on freshwater crayfish. Pages 23-46 in J. Kovonen and R. Lappalainen (eds.). *Rapautlous 2000, Keskis-Soumen Kalastuspiirin Kalastumoimisto Tiedous nro* 5, Jyväskylä, Finland.

Söderhäll, K., M. W. Johansson, and V. J. Smith. 1988. Internal Defence Mechanisms. Pages 213-235 in D. M. Holdich and T. S. Lowery (eds.). *Freshwater crayfish–biology, management and exploitation*.

Spitzy, R. 1973. Crayfish in Austria, history and actual situation. *Freshwater Crayfish* 1:10-14.

Strempel, K. L. 1975. Künstliche Erbrütung von Edelkrebsen in Zugergläsern und vergleichende Beobachtungen im Verhalten und Abwachs von Edel- und Signalkrebsen. *Freshwater Crayfish* 2:393-403.

Svärdson, G. 1965. The American crayfish *Pacifastacus leniusculus* (Dana) introduced into Sweden. *Reports of the Institute of Freshwater Research*, Drottningholm 46:90-94.

Svärdson, G., M. Fürst, and A. Fjälling. 1991. Population resilience of *Pacifastacus leniusculus* in Sweden. *Finnish Fisheries Research* 12:165-177.

Tarshis, I. B. 1978. Diets, equipment, and techniques for maintaining crawfish in the laboratory. *Proceedings of the World Mariculture Society* 9:259-269.

Tcherkashina, N. Y. 1977. Survival, growth, and feeding dynamics of juvenile crayfish (*Astacus leptodactylus cubanicus*) in ponds and the River Don. *Freshwater Crayfish* 3:95-100.

Thompson, A. G. 1990. Crayfish–a candidate for culture in British Columbia. *World Aquaculture* 21(2):92-94.

Unestam, T. 1969. Resistance to the crayfish plague in some American, Japanese, and European crayfishes. *Reports of the Institute for Freshwater Research*, Drottningholm 49:202-209.

Unestam, T. 1972. On the host range and origin of the crayfish plague fungus. *Reports of the Institute for Freshwater Research*, Drottningholm 52:192-198.

Unestam, T. 1973. Significance of diseases on freshwater crayfish. *Freshwater Crayfish* 1:135-150.

Unestam, T., K. Söderhäll, L. Nyhlén, E. Svensson, and R. Ajaxon. 1977. Specialization in crayfish defense and fungal aggressiveness upon crayfish plague infection. *Freshwater Crayfish* 3:321-331.

Vey, A. 1977. Studies on the pathology of crayfish under rearing conditions. *Freshwater Crayfish* 3:311-319.

Vey, A. 1979. Recherches sur une maladie des écrevisses due au parasite *Prosospermium haeckeli* Hilgendorf. *Freshwater Crayfish* 4:411-418.

Vey, A., and C. Vago. 1973. Protozoan and fungal diseases of *Austropotamobius pallipes* Lereboullet in France. *Freshwater Crayfish* 1:165-179.

Vey, A., K. Söderhäll, and R. Ajaxon. 1983. Susceptibility of *Orconectes limosus* Raf. to the crayfish plague *Aphanomyces astaci* Schikora. *Freshwater Crayfish* 5:284-291.

Vigneux, E. 1979. *Pacifastacus leniusculus* et *Astacus leptodactylus*. Premier bilan d'exploitation en etang. *Freshwater Crayfish* 4:227-234.

Walker, A. 1975. Crustacean aquaculture. *Proceedings of the Nutrition Society* 34:65-73.

Westin, L., and R. Gydemo. 1986. Influence of light and temperature on reproduction and moulting frequency of the crayfish, *Astacus astacus* L. *Aquaculture* 52:43-50.

Westman, K. 1973a. The population of the crayfish *Astacus astacus* in Finland and the introduction of the American crayfish *Pacifastacus leniusculus* Dana. *Freshwater Crayfish* 1:41-55.

Westman, K. 1973b. Cultivation of the American crayfish *Pacifastacus leniusculus*. *Freshwater Crayfish* 1:211-220.

Westman, K. 1985. Effects of habitat modification on freshwater crayfish. In J. S. Alabaster (ed.). Habitat modification and freshwater fisheries. Proceedings of a Symposium on Stock Enhancement in the Management of Freshwater Fisheries, in Budapest, Hungary, May-June 1982. *EIFAC Technical Paper.* (42): 21-24.

Westman, K., and V. Nylund. 1979. Crayfish plague, *Aphanomyces astaci*, observed in the European crayfish, *Astacus astacus*, in Pihlajavesi waterway in Finland. *Freshwater Crayfish* 4:419-426.

Westman, K., and V. Nylund. 1984. *Rapu ja ravustus*. Weilin and Göös, Espoo, Finland, 173 pp.

Westman, K., H. Ackefors, and V. Nylund. 1992. *Kräftor-Biologi, Odling, Fiske*. Kiviksgårdens Förlag. ISBN 91-971188-4-2. 152 pp.

Westman, K., M. Pursiainen, and P. Westman. 1990. Status of crayfish stocks, fisheries, diseases and culture in Europe. *Report of the FAO European Inland Fisheries Advisory Commission (EIFAC) working party on crayfish.* RKTL, Kalatutkimuksia 3, 1990. Helsinki, Finland.

Westman, K., M. Pursiainen, and R. Vilkman. 1979. A new folding trap model, which prevents crayfish from escaping. *Freshwater Crayfish* 4:235-242.

Wintersteiger, M. R. 1985. Zur Besiedlungsgeschichte und Verbreitung des Flusskrebse im Land Salzburg. *Österreichs Fischerei* 38:220-233.

Zandee, D. I. 1962. Lipid metabolism in *Astacus astacus* (L.). *Nature* 195: 814-815.

Zandee, D. I. 1966. Metabolism in the crayfish *Astacus astacus* (L.). III. Absence of cholesterol synthesis. *Archives Internationales de Physiologie et de Biochimie* 74:435-441.

Cultivation of Freshwater Crayfishes in Australia

B. J. Mills

Freshwater-crayfish Aquaculture Research and Management
Lymington, Tasmania, Australia

N. M. Morrissy

Western Australian Marine Research Laboratories
North Beach, Western Australia, Australia

J. V. Huner

Crawfish Research Center
University of Southwestern Louisiana
Lafayette, Louisiana, USA

Section I:
Introduction

Freshwater crayfish belonging to the family Parastacidae are those crayfishes inhabiting the Southern Hemisphere and differ from the Astacidae and Cambaridae of the Northern Hemisphere in the secondary sexual characters, especially those of the male. Parastacid crayfishes are recorded from South America (Riek 1971), New Zealand (Archey 1915), Madagascar (Holthuis 1964), Australia (Riek 1969), Papua New Guinea (Holthuis 1949), and the Aru Island (Roux 1914, 1933). There are no records of parastacid crayfish being collected from the African continent or the Indian subcontinent. Of the known number of species of parastacids in Australia (over 100) little is known of their biology or more importantly their aquaculture potential.

Until the early 1970s Australian aquaculture had been traditionally limited to pearl culture in northern Western Australia, the culture of Sydney rock oysters in New South Wales, and the freshwater
culture of rainbow trout in Victoria, New South Wales, and Tasmania. Since that time there has been considerable interest, research
effort, and commercial development of other species of finfish,
crustaceans, and molluscs for aquaculture in Australia. One crustacean group that has received considerable attention are the freshwater crayfish (Morrissy 1983d; Morrissy et al. 1990), particularly
over the last decade although production in 1990/91 was only 150
metric tons (Treadwell et al. 1991).

In this chapter the evolution, distribution, and the suitability of
the Australian crayfish fauna for aquaculture is briefly discussed.
Detailed information is provided on the biology and aquaculture
potential of three species of crayfish which are considered to hold
most promise for aquaculture. Finally the economic situation for
Australian crayfish farming is briefly discussed.

Section II:
Phylogeny

Riek (1972), in a study on the evolution of the group, suggests
parastacid crayfish can be divided into two ecological groups which
are correlated with morphological attributes.

The first group are strong burrowers which hold the chelae in a
vertical plane and this group comprise *Parastacus* in South America and *Engaeus*, *Engaewa*, and *Tenuibranchiurus* in Australia (Figure 69). All species are small in size, although *Parastacus* can reach
a moderate size. The combined branchial cardiac and post cervical
groove is well separated from the cervical grooves in most species
within the group, but are close in *Parastacus*. The strong burrowing
tendencies of these crayfish are correlated with the action of the
chelae and a reduction in the size of the abdomen (particularly in
the genus *Engaeus*). This group has a uniform structure of the male
genital papilla which is situated on the mesial projection of the
coxopodite. The projection is similar in texture and continuous with
the main calcified part of coxopodite, and the terminal apeture is
small. In *Tenuibranchiurus* the whole structure is very flattened.

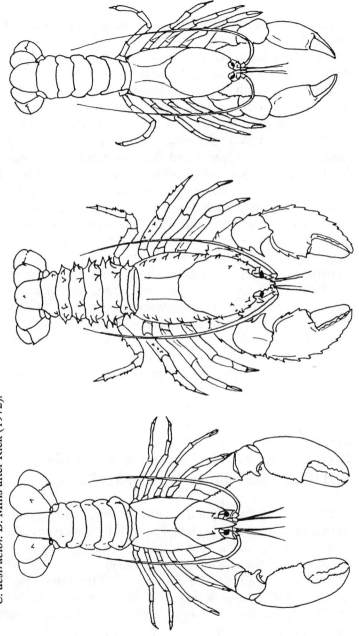

Figure 69. Left to right. *Engaewa subcoelrulea* representing the strong burrowing group of parastacid crayfishes. Note its small tail size. *Euastacus armatus* and *Cherax destructor*. These two species represent moderate burrowers. Note the spiny exoskeleton and small tail of *E. armatus* and the smooth exoskeleton and relatively large tail of *C. destructor*. B. Mills after Riek (1972).

219

The second ecological group are, in general, moderate burrowers, and contain the genera *Euastacus, Eustacoides, Astacopsis, Cherax, Geocharax, Gramastacus*, and *Parastacoides* in Australia, *Astacoides* in Madagascar, *Paranephrops* in New Zealand, and *Samastacus* in South America. These species are stoutly built and often reach a large size (e.g., *Astacopsis* and *Euastacus*). The separation of the post cervical and cervical grooves vary in this group from well separated (*Geocharax, Gramastacus, Paranephrops*) to close but distinct (*Cherax* and *Parastacoides*) and to not separable (*Euastacus, Eustacoides, Astacopsis*, and *Astacoides*). Within this group, several variations in the structure of the male genital papilla exist. A partial or complete sclerotised ring on the papilla separated from the calcified coxopodite may be present as in *Euastacus, Eustacoides, Astacopsis, Astacoides*, and *Samastacus*, or a moderate to large papilla without a hardened ring or tube may be present as in *Parastacoides, Cherax, Paranephrops*, and *Gramastacus*.

Although *Geocharax, Gramastacus*, and *Samastacus* are placed in the second moderate burrowing group, they differ in that the chelae tend to be held obliquely and have widely separate post cervical and cervical grooves, both of which are attributes of the first group. However, it is considered that these two characters are primitive and are attributes of the ancestral Parastacidae rather than of recent genera.

Section III:
Distribution of Parastacids in Australia and Their Suitability for Aquaculture

The general distribution of each Australian crayfish genera is shown in Figures 70, 71, and 72. To fully understand the distribution of parastacids in Australia, a knowledge of the climate and topography of the continent is necessary.

Australia has a land area of more than 7.6 million square kilometers, almost the same as the USA. The climate of Australia is predominantly continental with 50% of the area having a median rainfall of less than 300 mm per year and 80% with less than 60 mm. Extreme maxima, however, are encountered reaching 50°C over the inland. The generally low relief of Australia causes little obstruction to the atmospheric systems which control climate with the exception of the eastern uplands which modify atmospheric flow. In winter (May-October), anticyclones pass from west to east across the continent, at which time, northern Australia is influenced by mild, dry, southeast trade winds, while southern Australia is influenced by cool, moist, westerly winds. In summer (November-April), the anticyclones travel west to east on a more southerly track across the southern fringes of Australia directing easterly winds over the continent. Fine, warm weather predominates in southern Australia, while northern Australia is influenced by southern disturbances associated with the southward intrusion of warm, moist monsoonal air from north of the intertropical convergence zone resulting in a hot, rainy season. The mean annual rainfall across Australia is shown in Figure 73 and ranges from about 100 mm east of Lake Eyre to over 4000 mm near Cairns in Queensland. Howev-

Figure 70. Distribution of parastacid genera *Engaeus, Engaewa,* and *Tenuibranchiurus* in Australia. B. Mills after Riek (1969).

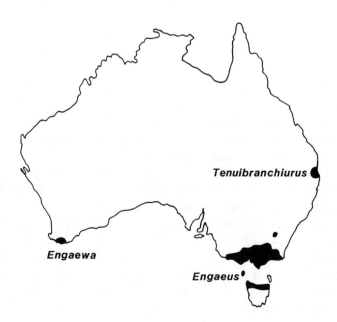

Figure 71. Distribution of parastacid genera *Astacopsis, Euastacoides, Euastacus, Geocharax,* and *Parastacoides* in Australia. B. Mills after Riek (1969).

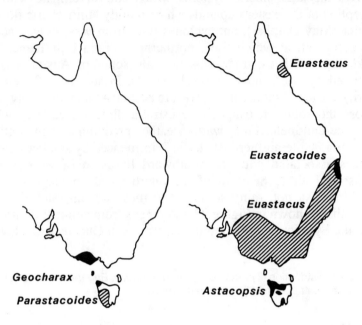

er, a large area of Australia receives less than 400 mm per year (Anonymous 1986).

The Australian freshwater crayfishes occupy most of the continent (excluding the arid central area) with the exception of the northwest coastal regions. This latter area is inhabited by the giant freshwater prawn, *Macrobrachium rosenbergii* and subspecies, which may exclude crayfish from this region.

For convenience of discussion of the aquaculture potential of Australian crayfishes, three broad categories have been established according to morphology of the body and size.

The first group is characterized by small body size (approximately 40 mm total length [TL]) and small tail size in comparison with the carapace. *Engaeus, Engaewa,* and *Tenuibranchiurus* comprise this group. There are approximately 23 species within the *Engaeus* genus and are recorded for most of the wet areas of Victoria, South Australia, and Tasmania. Maximum recorded size for an *Engaeus*

Figure 72. Distribution of the parastacid genus *Cherax* in Australia. B. Mills after Riek (1969).

Cherax spp.

species is about 80 mm TL. Only one species of *Tenuibranchiurus* occurs in Australia and this is recorded from southern Queensland. The size of the one specimen described was 25 mm TL. There are three species of *Engaewa* recorded from the wet areas of southwestern Western Australia. The maximum size recorded for a number of these species is about 40 mm TL.

In general, the small tail size of the species of *Engaeus*, *Engaewa*, and *Tenuibranchiurus*, their small maximum size, and their burrowing activities make them unsuitable for aquaculture.

The second group includes species which are characterized by a spiny exoskeleton and small tail-to-carapace size. The genera of this group are *Astacopsis*, *Euastacus*, and *Euastacoides*. A secondary characteristic is that many of the species grow to a very large size.

There are two recognized species of *Astacopsis* and they occur in the north, west, and south of Tasmania. Sizes range from approxi-

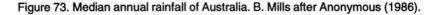

Figure 73. Median annual rainfall of Australia. B. Mills after Anonymous (1986).

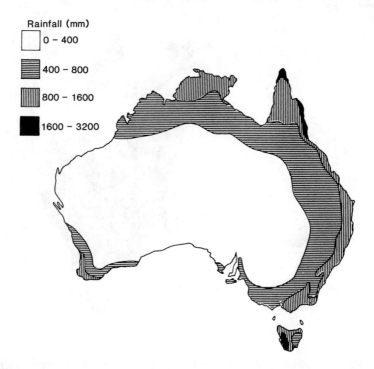

mately 50 to 400 mm TL with one species, *Astacopsis gouldii*, being the largest crayfish species in the world. Within the genus *Euastacus* seven species are recognized. Their distribution extends from southeast South Australia, Victoria, south and east New South Wales, and southern Queensland. An isolated pocket of a small *Euastacus* species also occurs in northern Queensland. Species within this group vary considerably in size, ranging from 65 mm to over 300 mm TL. The genus *Euastacoides* occurs in southern Queensland and comprises three known species. The maximum sizes recorded for a species of this genus is 70 mm TL.

The small size (and tail size) of *Euastacoides* would make species within this genera unsuitable for aquaculture. The larger species belonging to the genera *Euastacus* and *Astacopsis* would seem to be of interest to aquaculturists. Preliminary studies of the two larger species, *Euastacus amartus* (O'Connor, personal communication, 1986) and *Astacopsis gouldii* (Forteath 1985), seem to indicate that slow growth rates make both species unsuitable for aquaculture. Similarly with *Euastacoides*, both genera have a small tail size.

The third group consists of four genera which are characterized by a smooth exoskeleton (i.e., minimal spination) and relatively large tail compared with carapace size. The genera are *Geocharax*, *Parastacoides*, *Cherax*, and *Gramastacus*.

Only two species of *Geocharax* have been described, and their occurrence is restricted to northwest Tasmania, King Island, southwest Victoria, and southeast South Australia. Both species are small in size with a maximum known body length of 85 mm TL. Within the genus *Parastacoides* six species have been recorded. All species have a restricted distribution being confined to the west and southwest coast of Tasmania. The body length of any of the species rarely exceeds 60 mm TL. There are two species of *Gramastacus* which have a very restricted distribution in two small regions of Victoria and South Australia. Species are found in shallow swamps, are nonburrowing and rarely exceed 40 mm TL. The genus *Cherax*, comprising 27 species, is the most widespread genus within Australia occurring in all states (and climates) other than Tasmania. There is marked variation in size among the species, ranging from 60 to 400 mm TL.

The small maximum size of species belonging to the genera *Geocharax*, *Parastacoides*, and *Gramastacus* make them unsuitable for aquaculture. In addition, the species in the genera *Geocharax* have a strong tendency to burrow and inhabit marshland rather than permanent water bodies. Members of the genus *Cherax* that have larger maximum sizes (greater than 40 g) should be suitable for aquaculture. The largest *Cherax* species (2 kg), the marron (*Cherax tenuimanus*), and the most widely distributed species, the yabbie (*Cherax destructor*), would seem logical species for aquaculture. The yabbie has especially strong tolerances for environmental stresses. Another species of *Cherax* (*Cherax quadricarinatus*) has recently been shown to have considerable potential for aquaculture.

Section IV:
The Yabbie, *Cherax destructor*

TAXONOMY

The valid scientific name for the yabbie (Figure 74) is *Cherax destructor* (Clark 1936). Using only taxonomic references, the synonomy of *Cherax destructor* is as follows:

- *Astacoides bicarinates* (McCoy 1888)
- *Astacopsis bicarinates* (Haswell 1888, 1893; Spencer and Hall 1896)
- *Cheraps bicarinates* (Ortmann 1891, 1902; Faxon 1898)
- *Parachaeraps bicarinates* (Smith 1912; McCulloch 1914; Hale 1925, 1927)
- *Cherax destructor* (Clark 1936; Holthuis 1949, 1950; Riek 1951, 1956, 1969; Ziedler 1982; Sokol 1988)
- *Parachaeraps destructor* (Shipway 1951)
- *Cherax albidus* (Riek 1951; Hobbs 1974; Grant 1978; Sokol 1988)

Riek (1969) placed *Cherax destructor* in a group within the *Cherax* genus known as the "destructor group." Other members of this group are *Cherax albidus, Cherax davisi* (Clark 1936), and *Cherax esculus* (Riek 1956). Kane (1964) felt that *Cherax davisi* was synonymous with *Cherax destructor* but that *Cherax albidus* was a separate species. In his major study of the biology of the yabbie, Woodland (1967) names the species *Cherax albidus* but later renamed it *Cherax destructor* (Mills and Geddes 1980).

Using immunological techniques, Clark and Burnet (1942) investigated the antigenicity of the blood of three eastern Australian species, *Cherax destructor, Cherax albidus,* and *Cherax rotundus,* and two Western Australian species, *Cherax bicarinates* and *Cherax quinquecarinates.* They found that the three eastern species were more closely related to each other than were the two Western Australian species. Antigen tests with *Euastacus* and *Engaeus* suggested *Cherax* was closer to *Engaeus* than to *Euastacus.* Using both

Figure 74. Yabbie, *Cherax destructor*. B. Mills.

electrophoretic and immunochemical techniques, Patak (1982) and Patak and Baldwin (1986) investigated similarities of haemocyanin of six genera of Australian crayfish. In these studies, they found that the haemocyanin of *Cherax albidus* was indistinguishable from that of *Cherax destructor*, but both of these species were distinct from the Western Australian species, *Cherax plebejus* and *Cherax glaber*. Zeidler (1982) states that *Cherax destructor* was very difficult to distinguish from *Cherax albidus*. In a recent study Sokol (1988) stated there was no distinction between *Cherax destructor* and *Cherax albidus* based on morphological characters; and Austin (1986), in an unpublished study, considered the two species were not separate based on electrophoretic evidence.

DISTRIBUTION

The natural distribution of *Cherax destructor* has been recorded by numerous authors (Clark 1936; Riek 1951, 1969; Kane 1964; Woodland 1967; Reynolds 1980; Zeidler 1982; Sokol 1988). Introductions of yabbies have occurred into Western Australia

(Shipway 1951) and illegally into Tasmania, but with subsequent eradication (Ritchie 1978).

The yabbie has the most widespread distribution of all the Australian crayfish fauna. The species inhabits areas ranging from temporary lakes and rivers in hot, arid, central Australia to the cool permanent lakes and rivers of the southeastern highlands (Sokol 1988).

MORPHOLOGY AND ANATOMY

The yabbie is a moderately sized *Cherax* species reaching 220 g (A. Hudson, personal communication, 1987). The color of the exoskeleton can be variable-iridescent blue, off-white, orange, or pink. The most common color is, however, a light brown body (including the legs) with mottled light blue-green chelipeds. The major anatomical features of the exoskeleton are similar to other crayfishes. Several deformities of the exoskeleton have been noted.

Yabbies can be distinguished from other sympatric *Cherax* spp. by several characteristics. The sternal keel has four poorly defined keels on the dorsal surface of the carapace with a short smooth rostrum. There is a mat of fine hair on the upper, inner surface of the chelae. Yabbies have mottled mosaic pattern on the outer edge of the chelae. The amount of flesh in yabbies varies with size and sex (Reynolds 1980). Small yabbies (20-30 mm orbital carapace length [OCL]) have approximately 24% flesh in the abdomen, whereas large animals (70-80 mm OCL) have only 15%. Chelae flesh weight ranges from 4% in small animals to 8% in large animals.

LIFE HISTORY AND ECOLOGY

Within its range, yabbies occur in ponds, farm ponds, billabongs, swamps, creeks, rivers, bore drains, irrigation channels, roadside ditches, and reservoirs (Hale 1925; Clark 1936; Kane 1964; Woodland 1967; Riek 1969; Johnson 1978; Reynolds 1980; Lake and Sokol 1986). Horwitz and Richardson (1986) classify yabbies in permanent habitats as "Type 1b" burrowers. Such crayfish construct burrows on banks of permanent waters with openings above and below mean water level. In river environments, they preferentially inhabit flood plains (i.e., swamps, creeks, channels) rather than the main river channel. In the case of the River Murray the main channel is occupied preferentially by the large crayfish *Euastacus armatus* (Walker 1982, 1983; P. O'Conner, personal communication, 1987). Yabbies also inhabit temporary habitats such as farm dams, waterholes, creeks, and ditches which dry out in summertime (Clark 1936; Francois 1960; Reynolds 1980; Maher 1984). The yabbie is able to survive by burrowing to the water table or to where the soil is moist, returning to the land surface when surface water becomes re-established (Hale 1927; Frost 1975; Reynolds 1980). Dispersal of yabbies takes place at times of flood and/or wet nights where the yabbie is capable of travelling overground (Francois 1960; Frost 1975).

There are no published accounts of the distribution of yabbies in natural habitats. From observations of fishing patterns in Lake Alexandria, South Australia, and the Department of Fisheries' studies in the lake (B. Mills, unpublished data), certain areas were found to contain no yabbies, but in areas where yabbies occurred, an obvious preference for shoreline rather than open water was obvious (the limnological features of the lake have been discussed by Geddes [1984]).

Trapping exercises and location of burrows indicate that yabbies occupy the shallow areas of farm ponds (Woodland 1967; de Kretser 1979; Reynolds 1980). Lake and Sokol (1986), however, found no relationship between number trapped and depth, although juveniles were not sampled. de Kretser (1979) also suggested that distribution changed with season with yabbies moving into deeper waters in winter. Reynolds (1980) concluded, based on tagging studies, that a certain portion of the population moves continuously and has no real habitat preference.

There have been numerous published observations of the natural diet of the yabbie (Clark 1936; Woodland 1967; Anonymous 1979; de Kretser 1979; Reynolds 1980; Faragher 1983). Most authors have simply referred to food as mud, detritus, and plant material. Closer microscopic examination of the stomach contents (B. Mills, unpublished data; Anonymous 1979; Reynolds 1980; Faragher 1983) reveals that the diet has a variety of components. In general, material of plant origin dominates the diet with material of animal origin making up a relatively small proportion. Most individuals select animal material if available, however, in laboratory feeding trials, fresh plant material was consumed in preference to decaying plant material. In addition, adult yabbies were unable to capture many aquatic invertebrates (amphipods, shrimps, aquatic insect larvae) inhabiting their natural environment (B. Mills, unpublished data).

Growth rates of yabbies, like other crayfish species, depend upon such things as density, food availability, temperature, and other environmental conditions. In natural conditions growth occurs in the warmer months. Reynolds (1980) found that major molting (growth) occurred during the period from October to December, with minor growth occurring during the remainder of the growth period (January to March). Woodland (1967) and Lake and Sokol (1986) recorded molting throughout the spring/autumn period. Reynolds (1980) found reduced growth rates in breeding females and wounded yabbies. As with other crayfishes, growth rates decrease with increased size (Reynolds 1980; Lake and Sokol 1986).

There have been few studies to estimate growth in field populations. Woodland (1967), Reynolds (1980), and Lake and Sokol (1986) in their studies demonstrated that female growth slowed with onset of maturity. The average growth rate of juveniles in studies of pond populations by Woodland (1967) was 0.18 mm/day cephalothorax length (CL), while adult males grew 0.02-0.14 mm/day and females from 0.007-0.12 mm/day CL. Reynolds (1980) and Lake and Sokol (1986) recorded similar growth rates in their studies. In intensive culture conditions (see also Aquaculture section), Mills and McCloud (1983) recorded growth rates of up to 0.25 mm/day CL with some yabbies reaching market size (45 g) within six months.

Optimum temperature for growth is 28°C, with yabbies surviving up to about 36°C and ceasing to grow below 15°C (Mills 1986). The most suitable climate for this species would be Mediterranean-subtropical.

There is little published information on the ecology of yabbies in their natural habitats. Most studies have been conducted in man-made farm ponds. Woodland (1967) estimated total population numbers in a farm pond in a typical year, and found densities were high in early summer (20 yabbies/m^2) due to new recruits, and steadily declining to a low density in winter (2 yabbies/m^2) due to juvenile mortality. In a farm dam study, de Kretser (1979) found a population density of about 0.1-0.2 yabbies/m^2 for yabbies greater than 26 mm CL while Reynolds (1980) found 0.6-1.3 yabbies/m^2 for animals larger than 20 mm OCL. Lake and Sokol (1986) found farm pond populations varied in density from 0.6-1.0 yabbies/m^2.

Data presented by Woodland (1967) suggests that recruitment of young third instar yabbies into the population is relatively high at about 18,000 juveniles/100 breeding females. Mortality was found to be higher in winter in contrast with de Kretser's (1979) findings which concluded that mortality was highest in the summer months.

Standing crop estimates made by Woodland (1967) and Reynolds (1980) for farm ponds have fairly close agreement at 340 kg/ha and 274 to 332 kg/ha, respectively. de Kretser's (1979) study found a standing crop of only 77 kg/ha, while Lake and Sokol (1986) estimated their farm ponds to have standing crops ranging from 180 to 270 kg/ha.

Woodland (1967) found that yabbies had a net growth efficiency of 54% (ranges for other crayfish, 20-30% [Momot 1984]). Lewis (1976) found that both gross production efficiency and net growth efficiency increased with increased temperature. Net production was found to be influenced by feeding rates and not stocking rates in experiments conducted by Mills and McCloud (1983).

There is little information on tolerance ranges to water quality for the yabbie. Dissolved oxygen is the prime concern in survival of yabbies. The survival time for an LC$_{50}$ of 4.9% oxygen saturation is 45 hours compared with 14 hours for marron (Lewis 1976). The critical oxygen concentration for yabbies is 2.7 mg oxygen/liter (Barley 1983). There have been no studies of the effect of low oxygen on growth of yabbies.

In addition to oxygen tolerances, Mills and Geddes (1980) found that juveniles were less tolerant to salinity than adults with 96-hour LC$_{50}$ values of 25.8 parts per thousand (ppt) and 29.9 ppt, respectively. Mills and Geddes (1980) found that growth of yabbies was affected at about 6 ppt salinity. There is no information on tolerances to pH and calcium, but they are likely to be similar to those of other crayfish species. Tolerances of yabbies to pollutants, princi-

pally heavy metals and pesticides, have also received little attention. Skidmore and Firth (1983) found a 96-hour LC_{50} concentration of copper at 1.4 ppm. There is no published information on the
toxicity of pesticides to yabbies.

The sex of yabbies, can be distinguished by the position of the
genital apertures: in males, the genital apertures occur on the coxopodite of the fourth walking legs (5th periopods) and on the coxopodite of the second walking leg (3rd periopods) in females (Hale
1927; Woodland 1967). The male genitals are a pair of long complex
papillae while the genitalia of the female consist of a pair of oval
openings covered by a noncalcified membrane (Johnson 1979). Male
yabbies attain a larger size, have larger chelipeds and narrower abdomens compared with females. When females attain sexual maturity,
they develop filamentous setae on the pleopods for egg attachment
(Woodland 1967; Lewis 1976). In all other respects the sexes are
similar. Ovarian development has been discussed by Johnson (1979).

Riek (1951, 1956), Woodland (1967), Johnson (1979), Reynolds
(1980), and Lake and Sokol (1986) recorded the occurrence of
intersexes in Australian parastacid crayfishes. In farm dam studies
by Johnson (1979) several percent of a population had intersexes.
Dissection of specimens revealed either functional males or functional females. The most common form noted was that of three
openings. In some cases, males with ovarian tissue were found and
in one case, a female was found to have testicular tissue in its ovary.
In one case, Johnson (1979) found a true hermaphrodite having
normal testes on the right and normal ovary on the left, with a vas
deferens and oviduct leading to the gonopores.

The breeding season for yabbies occurs in the summer months,
but the duration of breeding will depend on the geographic location,
e.g., from September to March in western New South Wales (Reynolds 1980) and from December to March in southeastern New South
Wales (Faragher 1983), and from September to May in the wheat
belt region of Western Australia (Morrissy, unpublished data).

Females naturally breed only once per year (Woodland 1967)
although Johnson (1979), Reynolds (1980), and Lake and Sokol
(1986) believe that females are capable of breeding at least twice
within the breeding season. Regarding only incubation times at summer temperatures, it would be possible to have 2-3 breedings per

season (Mills 1986). Woodland (1967) and Lewis (1976) suggest that increasing water temperatures is the major cue for mating and spawning; however, Johnson (1979) felt that although temperature was an important stimulus for spawning, photoperiod was the major cue. Mills (1986) (and see Aquaculture section) found photoperiod to be the major cue for out-of-season breeding in yabbies.

There are no records of the proportion of females in each year of life that spawn. However, Reynolds (1980) found that approximately 40% of the females in mature condition spawn in September and October with between 2-10% spawning in the other months of the spawning season. The minimum size for females capable of spawning ranges from 26 mm OCL (Faragher 1983) to 38 mm OCL (Johnson 1979).

The number of juveniles produced by crayfish at hatching is generally less than ovarian egg count (see, for example, Morrissy 1974). In yabbies, however, the majority of eggs within the ovary are spawned at the time of mating (Johnson 1979). The number of spawned eggs lost due to disease, mechanical damage, and disturbance (at least from aquarium observations) is minimal (B. Mills, unpublished data). Actual fecundity, therefore, is very closely related to potential fecundity in this species. Fecundity is directly related to female size (Figure 75).

Both Woodland (1967) and de Kretser (1979) found sex ratios that favored females. Reynolds (1980) found an overall 1:1 ratio for her study in ponds while Lake and Sokol (1986) found some ponds had 1:1 ratios but others favored one or the other sex. In a study of a natural population in Lake Alexandria (B. Mills, unpublished data), it was found that sex ratios ranged between 18:1 to 0.4:1. As pointed out by Lake and Sokol (1986) a bias exists in catchability of sexes not only for brooding females but also for size domination especially affecting the smaller size classes in trap catches. In a study of newly hatched juveniles from laboratory rearing, Mills (1980) found a ratio of 2:1.

PREDATORS AND COMPETITORS

Predators of *Cherax destructor* are other yabbies, fish, and aquatic birds. Native fish eat yabbies and include Murray cod (*Maccullo-*

chella peeli) (Reynolds 1976; Cadwallader and Backhouse 1983), catfish (*Tandanus tandanus*) (Davis 1977), and golden perch (*Macquaria ambigue*) (B. Mills, unpublished data). Introduced fish species which consume yabbies include brown trout (*Salmo trutta*) and rainbow trout (*Oncorhynchus mykiss*) (Morrissy 1967; Pidgeon 1981; Faragher 1983), English perch (*Perca fluviatilis*) (Roughley 1966), goldfish (*Carassius auratus*) (Johnson 1979), tench (*Tinca tinca*) (de Kretser 1979), and possibly European carp (*Cyprinus carpio*) (Hume et al. 1983). Mosquitofish (*Gambusia affinis*) have been observed to consume juvenile yabbies (Carroll 1981).

Since carp and tench are primarily detritivores, it has been suggested that they may compete with yabbies for food (Reynolds 1976; Anonymous 1977, 1979). However, in most cases, detritus is not limiting and competition for food is unlikely.

Aquatic birds which prey on yabbies are the Little Pied Cormo-

Figure 75.The relationship between female yabbie size and the number of juveniles hatched.

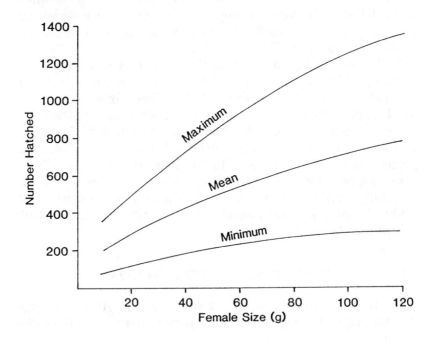

rant (*Phalacrocorax melanoleucos*) (McNally 1957; Woodland 1967; Miller 1976, 1979; de Kretser 1979; Johnson 1979), the Little Black Cormorant (*Phalacrocorax sulcirostris*), the Black Cormorant (*Phalacrocorax carbo*) (Grant 1978), and the Straw-necked Ibis (*Threskiornis spinicollis*) (Carrick 1959).

For the three species of turtles that inhabit the Murray River system yabbies comprised a major part of the diet of *Chelodina expansa*. They were minor components of the diets of *Chelodina longicollis* and *Emydura macquarii* (Chessman 1983, 1984).

The Eastern Water Rat (*Hydromys chrysogaster*) is a known predator of yabbies (Troughton 1941; Fleay 1964; Woollard et al. 1978; Shearer 1981). Both the introduced fox (Croft and Horne 1978) and the native dog (dingo) (Newsome et al. 1983) are predators of crayfish.

Yabbies were an important component of the diet of the native aboriginal people living in the Murray-Darling Basin (Eyre 1845; Kefous 1981a,b). More recently, European immigrants have become significant predators of this species both for food and fish bait.

PARASITES AND DISEASES

There is little information about diseases that affect Australian crayfishes. Considerable studies have been made of the diseases which affect Northern Hemisphere crayfish species (Johnson 1977). Mills (1983) reviewed known and potential diseases that affect yabbies.

There are several diseases that are known to affect yabbies. One common disease is a bacterial infection of the exoskeleton which causes it to be eaten away where damage occurs and is caused mainly by *Pseudomonas* sp. Protozoan diseases caused by the microsporidian *Thelohania* sp. have been found in populations of yabbies in Australia. Other protozoans are found on the gills of body surfaces of the yabbie but it is unknown whether these have any detrimental effects on health.

There are numerous parasites and commensals that occur on yabbies including species of rotifers, platyhelminthes, nematodes, and cestodes, mites, and crustaceans (ostracods). With exception of platyhelminthes (Figures 76A and 76B), few or no studies have

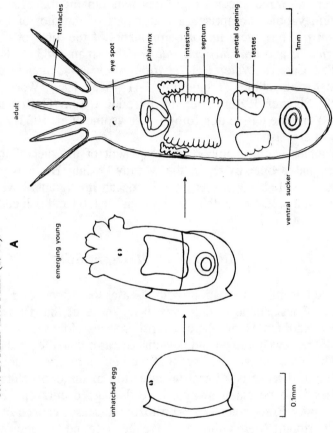

Figure 76. Examples of temnocephalid "flat worms" inhabiting yabbies (*Cherax destructor*) (A) *Temnocephala dendyi*, and (B) *Craspedella spenceri*. B. Mills after Mills (1983).

236

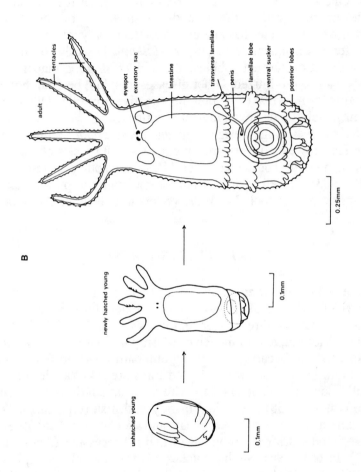

B

adult

tentacles

eyespot
excretory sac

intestine

transverse lamellae

penis

lamellae lobe

ventral sucker

posterior lobes

0.25mm

newly hatched young

0.1mm

unhatched young

0.1mm

237

been made of the species or their effects on yabbies although most are probably benign commensals.

Two other diseases of unknown cause have been noted in yabbies. One disease appears as warts or tumors on the exoskeleton while the other causes the uropods and telson to blister. It is unknown whether either has any adverse affects on yabbies.

In terms of aquaculture, only *Thelohania* is likely to have any major effect on farmed yabbies, although, with good management, effects can be minimized. Other diseases, may not be fatal, but their appearance may make yabbies unattractive for sale. Others may impair metabolic or physiological functions.

The crayfish fungus plague, *Aphanomyces astaci*, is unknown in Australia. However, unless proven otherwise, all species of parastacid crayfish must be considered to be highly susceptible to this disease (Mills 1983). Therefore, it is of utmost importance to prevent non-native species which could be plague vectors from entering Australia.

NATURAL FISHERIES

Over its natural range commercial fishing for yabbies occurs in large river systems, lakes, and to a lesser extent farm ponds. Fishing is concentrated in three major regions, western New South Wales, western Victoria, and in the River Murray in South Australia. In all regions fishing is carried out using outboard powered fiberglass or aluminum boats from which baited traps are set. Yabbie traps are usually constructed of chicken wire with inverted cone entrances (Anonymous 1968) and baited mainly with fish (e.g., carp). Traps are generally checked once a day and catches are held in small freezer stores prior to shipment to major processors (Anonymous 1973). Prior to 1968, yabbie catches were sold locally (e.g., hotels) but with the development of export markets most of the catch has been sent overseas, in particular to Sweden (Anonymous 1968; Unestam 1975).

In each state fishermen are required to furnish catch returns to departments responsible for fisheries. In South Australia an additional requirement is that effort figures (trap lifts) are to be provided to the Fisheries Department.

Catches by state show wide fluctuations over the years with a trend in reaching a peak followed by a period of decline (Table 22). Interpretation of the declining catches over the years is difficult due to the widespread nature of fishing activities (i.e., the fishery cannot be treated as unit stock) and in the case of two states (New South Wales and Victoria), no effort figures are available so that catch and effort analysis cannot be applied. The fluctuation in catches has caused concern with continuity of supply to markets, particularly the export market. This lack of continuity in supply of yabbies caused a cessation of export to Sweden in 1977 after a successful export period in the early 1970s.

It is unknown whether the apparent decline in catches was the result of overfishing or the result of some environmental factor. Since catch and effort figures are only available for the South Australian fishery any effect of overfishing in New South Wales and Victoria cannot easily be determined. For South Australia, catch and effort data from the major fishery area, Lake Alexandrina, has been analyzed (B. Mills, unpublished data) (Figures 77A, 77B). The figures show a gradual increase in catch from 1968 to a peak in 1974 of 80 tons after which a rapid decline occurred within the space of two years stabilizing at 10-20 tons annually. The catch per unit effort (CPUE) fluctuated markedly over the years but averaged 0.3 kg/potlift. The reason for the decline in catches after 1974 in uncertain. CPUEs show a slight decline with time which may suggest fishing pressure was a contributing factor to the catch reduction. Predators and competitors are limited to a few bird and fish species, but it is unlikely that they would have significant effect on yabbie numbers. Environmental influences likely to cause population decline are pollutants, salinity, dissolved oxygen, temperature, and flooding. With the exception of flooding, none of these variables exceed levels at which detrimental effects would influence the yabbie populations (Geddes 1984; B. Mills, unpublished data). From a comparison of catch rates and floods, it appears that until the collapse of the fishery, high catch rates were correlated with high water levels; however, after the collapse of the fishery, there has been no such correlation.

There have been no reports of disease outbreaks in the yabbie fishery which would contribute to the decline. The effect seems to

Table 22. Commercial fishery production of yabbies by State and Australia (tonnes): 1974/75 to 1990 (modified from Staniford et al. 1987)

Year	New South Wales	Victoria	South Australia	Total Australia
1974/75	16	2	127	145
1975/76	25	—	86	111
1976/77	26	5	44	75
1977/78	7	—	21	28
1978/79	16	19	3	38
1979/80	18	42	8	68
1980/81	7	61	4	72
1981/82	15	127	15	157
1982/83	2	61	7	70
1983/84	n.a.	9	10	n.a.
1984/85	n.a.	15	8	n.a.
1985-90	No significant commercial production recorded			

Source: Australia Bureau of Statistics, Fisheries, Australia, Catalogue No. 7603, various issues. Department Conservation Forests and Lands, Victoria (1984)

be permanent because catch rates have been unable to recover from the dramatic decline in catches after more than a decade. Communication with fishermen suggests that similar dramatic declines in yabbie catches have occurred in the two other states.

It would appear that if commercial exploitation of natural yabbie populations leads to uncontrolled rapid collapses in the fishery, then aquaculture may be a viable alternative to supply the market.

AQUACULTURE OF YABBIES

Yabbies can be cultured either extensively in large manmade lakes (in the range of 1-5 ha in size) or in natural lakes where little

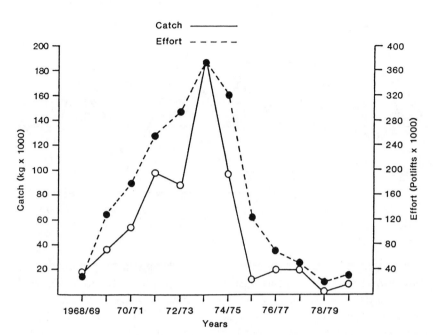

or no management is practiced, or semi-intensively in small ponds (units of about 0.1-1.0 ha in size). The first system is similar to that successfully practiced in crayfish farming in the USA, particularly in Louisiana (see, for example, Avault and Huner 1985). In Australia, this type of culture is limited by lack of large areas of inexpensive suitable land and access to inexpensive supplies of water. Limited production using this system is possible on farm land where there are numerous water supply dams.

In recent years there has been considerable interest in utilizing farm ponds for yabbie culture particularly in the wheat belt area of Western Australia where there are numerous farm ponds (0.1-0.2 ha). Approximately 3200 farm dams are currently used to culture yabbies and produce around 35 tons per year at an average of 50 kg per dam (Anonymous 1990). The basic method is based on either stocking hatchery-produced juveniles into ponds at low levels (less than $5/m^2$) into existing dams or ponds specifically constructed for cul-

Figure 77B. Catch per unit effort for the South Australia yabbie fishery.

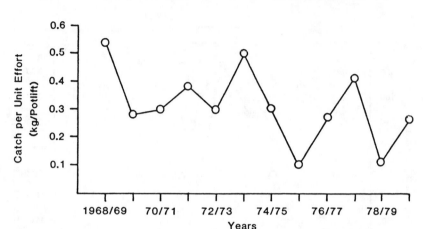

ture, or to stock adults. In either case stocking is aimed at producing sustained populations through natural reproduction. Food for yabbies consists of naturally occurring feed, mainly detritus, but in some cases supplementary feed is supplied (usually chicken or lucerne pellets) where higher production rates are desired. Management of ponds is minimal, in some cases predator control, supplementary feeding, controlled harvesting, and environmental monitoring is exercised. Production rates are low, around 300 kg/ha, but where there is a degree of pond management production of 1000 kg/ha may be achieved.

The second system using semi-intensive culture practices consists of three distinct production phases, namely, juvenile production (hatchery), juvenile growout (nursery), and adult growout (Geddes 1984; Momot 1984). The hatchery phase has been fully researched and is being used commercially. The nursery and growout phases are still at the research stage, although some commercial development has taken place.

To stock ponds on a year round basis, continual supply of juveniles from a hatchery is necessary. Under natural conditions, as previously noted, spawning only occurs during the summer months with a maximum of two spawnings (Johnson 1979; Mills 1986). The number of juveniles produced by a single, large female can be as high as 1200. Selection of large female breeding stock is, therefore, a primary requirement of efficient hatchery production. In addition, it would be desirable to select the most docile broodstock available through observation. Broodstock should also be disease free.

Provision of a photoperiod of 14L:10D or above using artificial light and a water temperature exceeding 18°C results in year round breeding. Higher temperatures can be used and lead to a reduction in incubation time; for example, at 20°C incubation time is 40 days whereas at 30°C the incubation time is reduced to 19 days; however, higher temperatures can lead to poor water quality and to fungal infections of the eggs if adequate water quality is not maintained.

The hatchery can be of simple construction, ranging from "garage" size with static water conditions and individual tank heating using aquarium immersion heaters, to a large scale multi-tank system with recirculated heated and filtered water that is recirculated. Spawning tanks can be made of glass, plastic or polystyrene (approximate dimensions 400 mm × 300 mm × 200 mm deep) which is large enough to house 2-3 females and one male. The water is aerated via air diffusers and in the static tank system fitting of undergravel filters reduces the number of water changes which normally occur at intervals of two weeks. Additional large tanks can be used with numerous males and females, but as yet this system has not been fully evaluated (Mitchell and Collins 1989).

Feeding of broodstock is best achieved using pellet feeds such as trout or chicken pellets. Feeding one pellet per individual every second day appears to maintain good stock condition (B. Mills, unpublished data), but higher feeding rates may increase general stock health although requiring greater water exchange to maintain water quality.

As previously mentioned, yabbies are capable of repetitive spawning which is an advantage in hatchery production of juveniles. Females vary in the number of young produced for a given size, and in the length of time between each spawning after hatching has occurred (range 10 days to 1 year). High mating success is of prime consideration in the efficient running of a hatchery (B. Mills, unpublished data) therefore selection of stock with average fecundity and minimum time between successive spawnings is essential.

Once the young leave the female they are transferred from the hatchery to outdoor juvenile nursery ponds. These ponds can be aboveground concrete or swimming pool types or earth in-ground with a hard bottom substrate and can be of any size. The ponds are covered with a fine netting to prevent predatory insects entering, the most damaging of which are dragonfly nymphs, dytiscid beetles, and notonectids. The ponds are provided with a substrate of high organic mud approximately 25 mm deep which appears to provide an adequate food source for juvenile yabbies (B. Mills, unpublished data) in the size range 15 mg to 5 g. The water in the pond should be aerated and plastic mesh bags (onion bags) or rope tassels suspended in the water column to increase the surface area for yabbies to occupy, and allow high stocking rates. Stocking rates in the range 100-300/m^2 have been used experimentally with growth to 1 g and 90% survival achieved within 6 weeks (Geddes et al. 1993).

There are several feed types for yabbies at the nursery stage and can include natural detritus, pellets, and live feed. In growth studies of yabbies in small pools with a detrital substrate it has been shown that newly hatched yabbies can grow to 1-5 g within 6 weeks at low stocking rates (10-30/m^2) (Mills and McCloud 1983; B. Mills, unpublished data). Recently, Mitchell and Collins (1989) have found that feeding zooplankton (Daphnia) has provided excellent growth and survival at stocking densities of 880/m^2. In growth trials with juvenile yabbies Smallridge et al. (1989) have shown that *Artemia* cysts and *Artemia* nauplii elicit good growth, although the feeding of live feeds may not be economically viable since yabbies are benthic feeders. Where zooplankton is used as feed it would seem preferable to use the frozen form which sinks to the bottom and is readily available to the crayfish (Jones 1989). Harvesting of advanced juveniles is best achieved by pond draining or less effectively by seine netting.

Once juveniles reach 0.5-1.0 g they are suitable for stocking in growout ponds because they are much hardier than newly hatched juveniles and are less cannibalistic (B. Mills, unpublished data).

After completion of the nursery phase juveniles are stocked into growout ponds (Figure 78). Earthen ponds ranging in size from 0.1 to over 1 ha are filled with water to a depth of between 1 and 2 m. Generally no refuges are provided and aeration is seldom used.

Stocking rates employed are in the range of 2-20 yabbies/m^2. Although growout of yabbies has been practiced for a number of years, aquaculture is still considered to be semi-commercial (B. Mills, unpublished data) with production being approximately 97 tons in 1990/91 (Treadwell et al. 1991). There are no published accounts of the methods used and yields obtained in commercial yabbie farms.

Experimental and semi-commercial trials have been carried out and a description of the methods follows.

In pilot experiments designed to determine optimum stocking and feeding rates for yabbies in growout ponds, Mills and McCloud (1983) stocked newly hatched juvenile yabbies in ponds at 10, 20, 30/m^2, and fed them lucerne at various rates. The result of these experiments showed that as stocking rate increased growth rates decreased due to stunting and as feeding rate increased growth rate increased. Growth appeared to be proportional to the amount of

Figure 78. Newly drained yabbie pond. Note large numbers of yabbies dispersed about the muddy bottom. B. Mills.

feed added. Individual growth was highly variable ranging from 15 to 80 g, with a mean size of 40 g (minimum market size) being reached in six months. Survival was independent of feeding rate but decreased with increased stocking rate. The final stable survival was 5 yabbies/m^2. Biomass (production) values reached the equivalent of 1500 kg/ha but were still increasing sharply at the end of the experiment indicating higher values could be achieved with longer growout times. Food conversion ratios, using lucerne, were less than 5:1. The conclusion derived from this study was that stocking at between 5-10/m^2 and feeding lucerne at 3 g/m^2 with an additional 3 g added each week will give good growth and production with reasonable survival rates in excess of 75%.

Mitchell and Collins (1989) in trials using small tanks stocked and fed yabbies at rates similar to Mills and McCloud (1983) and found similar growth and production rates. In semi-commercial production trials (Penberthy, personal communication, 1989) stocked hatchery-produced juveniles at 20/m^2 in three 0.3 ha ponds and fed them pelletized feeds. After approximately six months the ponds were drained and yielded approximately 1200 kg/ha of market-size yabbies (> 40 g) with the weight of approximately one third falling below market size.

Other experiments have indicated that other formulated stock feeds can double growth rates compared with lucerne (B. Mills, unpublished data), and in one case, pig feed costs about the same as lucerne. In addition, growth rates can be enhanced with planting of grass in ponds prior to stocking to provide an immediate food source for juveniles as well as providing a highly organic substrate after decomposition necessary for detritus formation (Maher 1984). Further research into nutrition at the growout stage is needed to optimize yields (Mitchell and Collins 1989).

Harvesting yabbies from growout ponds can be carried out by using seines, lift nets, or traps. Harvesting using these methods is important to remove fast-growing individuals (Mills and McCloud 1983). "Complete" harvest is accomplished by draining of the pond.

Until large-scale commercial production of yabbies is undertaken, evaluation of techniques described here remain uncertain particularly in the nursery, growout, and harvesting phases.

Section V:
The Marron, *Cherax tenuimanus*

TAXONOMY

The valid scientific name for marron (Figure 79) is *Cherax tenuimanus* (Clark 1936). Using only taxonomic references, the synomy of *Cherax tenuimanus* is as follows:

- *Cheraps tenuimanus* (Smith 1912)
- *Cherax tenuimanus* (McCulloch 1914)
- *Cherax tenuimanus* (Clark 1936)
- *Cheraps tenuimanus* (Shipway 1951)

DISTRIBUTION

The marron is native to southwestern Australia in the state of Western Australia (Figure 80). The original range included all of the streams of the Darling Range extending between the Harvey River and the Kent River. As a consequence of transplantations, they are now found in natural streams as far north as the Hutt River north of Geraldton, and southeastwards to the Esperance area. Marron have also been widely introduced into farm and water supply dams within and outside their native ranges in Western Australia with the extension of their range inland between Geraldton and Esperance, extending inland to Kalgoorlie and Lake Grace. Morrissy (1978a) has detailed information about the distribution of marron in Western Australia.

Within the last decade, restrictions on export of live marron from Western Australia have been lifted and there have been extensive transplantations to sites of extensive and intensive aquaculture in eastern Australia as well as overseas. There have been few reported successful transplantations of marron to any natural waters anywhere outside of Western Australia, with the exception of a successful transplantation on Kangaroo Island in South Australia (B. Mills, unpublished data).

Figure 79. Large marron, *Cherax tenuimanus*. J. Huner.

MORPHOLOGY AND ANATOMY

The marron is a large crayfish with specimens weighing up to 2.72 kg and measuring 385 mm TL, but these are exceptions. Legal size from natural fisheries is 76.2 mm OCL which corresponds to a weight of about 120 g. Marron are amongst the largest of the freshwater crayfish but species of the genera *Astacopsis* (Tasmania) and *Euastacus* (eastern Australian States) do grow larger (see previous discussion).

Marron are distinguished from other sympatric *Cherax* species

Figure 80. Natural distribution of yabbie, marron, and redclaw.

by several characteristics. There are five ridges, or keels, on the rostrum. There are two spines located centrally on the upper surface of the telson. There is no mat of fine setae on the upper surface of the chelae as is often found in other *Cherax* species and the chelae are also relatively narrow. Three pairs of distinct spines are present on the rostrum. The tail including the exoskeleton accounts for approximately 42% of the animal's total weight and provides the greatest meat-body recovery percentage of any widely exploited crayfish in the world.

Mature marron are brown-black. First year juveniles have dark markings over a yellow-green or brown background, while older juveniles are brownish. Bright blue and red color morphs have been reported.

Further information about marron morphology and anatomy can be found in Smith (1912), Morrissy (1976a), Olszewski (1980), Crook (1981).

LIFE HISTORY AND ECOLOGY

Marron prefer deep, broad waters or permanent streams, and are generally found in larger pools in their lower reaches (Morrissy 1978a, b). Marron are poor burrowers and do not survive at all if their habitat dries out. Horwitz and Richardson (1986) classify marron as "Type 1a: Burrows in Permanent Waters." Such crayfish are found in permanent bodies of surface water such as lakes, rivers, and large creeks occupying spaces under rocks, ledges, in rock crevices, in or under submerged logs, and in short, unbranched burrows in the substratum. In natural, clear water habitats marron usually rest during daylight hours and begin to forage for food at sunset. Population densities are closely related to the amount of shelter or debris available. Such waters may be classified as oligotrophic but standing stocks may vary from 300-600 kg/ha in favorable habitats.

Marron are omnivorous but, as with other freshwater crayfish, they depend heavily on microbially enriched plant detritus for basic sustenance (Morrissy 1974). Morrissy (1980) examined the stomach contents of marron from farm dams in the Wheatbelt area of Western Australia and found they fed largely on pond sediment detritus. He noted that organic content of stomachs was 81% while that of sediments was 11% implying that selective feeding occurred. Organic contents of stomachs decreased with increased size suggesting that larger marron were less adept at selecting finer particles with greater microbial enrichment. There was a strong negative correlation between benthic and planktonic crustaceans, molluscs, and insects and marron density. This observation indicates the possibility of competitive exclusion as well as direct predation.

Marron growth rates are variable, and depend on such factors as density, water temperature, food, and availability of cover/shelter. Individuals having a maximum weight of 2.72 kg are at least 5 years of age. Under experimental conditions, Morrissy (1979) found that at

initial population densities of 2-15/m² young of the year (0+ age
class marron)/m², a multiple regression equation accounted for
93.3% of the variability in log growth rate, with 45.7% due to initial
size, 40.6% due to seasonal water temperatures, 6.1% due to initial
density, and 0.8% due to feeding rate. A stocking rate of 5 young of
the year /m² would result in average growth to 45 g after one year
with 50% survival while the survivors would grow to 120 g at the
end of the second year of life, again with 50% survival. Variation in
growth rates of same age marron can be great (Crook 1981); for
example, a marron of 76.2 mm CL (55 mm OCL) and a weight of
120 g may be between 11.5 months and 5 years of age. Growth of
most marron to market size occurs between 6 and 18 months of age.

Optimum average temperatures for growth under natural conditions are in the 15-20°C range. Growth ceases below 12°C while
brief exposures to temperatures above 31.5°C are fatal. The optimal
temperature for growth, derived experimentally, is 24°C (Morrissy
1990). Morrissy (1976a, 1980) suggests that the most suitable natural range climate for marron culture is of the "Mediterranean" type.

Natural riverine populations of marron may reach standing crop
of 300-600 kg/ha (Morrissy 1978a, b). Populations can reach 1000+
kg/ha in earthen farm ponds and 3000+ kg/ha in earthen, intensively
managed culture ponds (Morrissy 1979, 1980, 1986).

Dissolved oxygen is critical consideration in marron survival.
Survival for an LC 50 of 4.9% oxygen saturation at 20°C was about
14 hours for marron compared to 45 hours for the hardier yabbie
(*Cherax destructor*) (Morrissy et al. 1984). Marron will die at oxygen concentrations lower than 40% of saturation at 20°C and
growth is poor if oxygen concentration is less than 70% saturation
(Morrissy 1976a). Morrissy (1978a) attributes the decline of natural
stocks of marron in many rivers to rapid eutrophication with con-
comitant reduction in oxygen levels and siltation arising from forest
exploitation and agricultural practices in the catchment basins of the
rivers. He notes, in fact, the sensitivity of marron stocks to the
depleted oxygen conditions accompanying eutrophication suggests
that marron could be a valuable "indicator" species for monitoring
this unfavorable by-production of development.

Recorded ranges of water quality tolerance for marron are wide:
salinity (conductivity as ohms per cm) 125-11,600, saline inland

waters for calcium ions, 1.2-148 ppm, and pH 7.0-8.5 for ponds (Morrissy 1976a). Marron are, however, especially sensitive to pesticides and heavy metals in the environment. Young of the year marron are referred to as 0+ age class animals. Older marron are designated as members of the 1+, 2+, etc., age classes. Morrissy (1974, 1976b, 1978a, 1983a, 1986) describes the reproductive cycle for marron with ovarian development commencing in January and February and proceeding until July when maximum ova size has been reached. Ovarian development (Morrissy 1974) appears to be initiated by decreasing day length, commencing soon after maximum day length in late December and completed by minimum day length in late June. Mating occurs in September being apparently stimulated by the start of the spring rise in water temperature. Mating can be induced as early as mid-July by early elevation of water temperature, but is aborted by long (summer) day lengths (Morrissy 1983b). Berried females appear in October and November. Females carry attached young in December and newly released juvenile marron are present in January. Up to 39% of the larger females of an age class may spawn during their second year of life but most, usually about 75%, of the females spawn in their third year of life.

The number of living young produced by a female marron is lower than the potential number based on ovarian egg counts. This is a normal situation for most crustaceans all eggs of which may not be extruded (Morrissy 1974). Diseases may destroy eggs, eggs may be lost as a result of mechanical disturbance, and/or there may be inadequate space on the pleopods for attachment of all eggs. The actual fecundity is much more important than potential fecundity for obvious reasons. Fecundity is directly related to size and the number of young that can be expected to be produced by a healthy female is shown in Figure 81.

There is direct correlation between food supply and ovarian development (Morrissy 1978b). In the absence of adequate nutrition, the female will resorb her eggs. Comparison of several wild marron populations ranging in density from 0.14 to 1.50/m^2 showed a negative relationship between density and growth rate and size at maturity. Smallest size of spawning females varied from 2.8 to 7.3 cm OCL. About 79% of the females from the dense, slow growing populations were "protected" from exploitation being smaller than

Figure 81. The relationship between female marron size and the number of juveniles hatched.

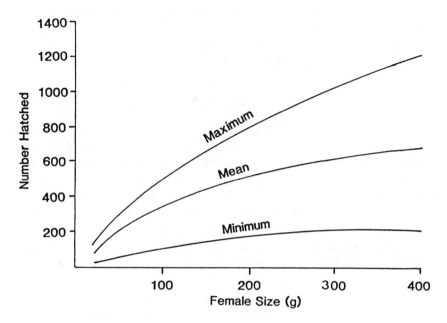

the legal sizes of 5.5 OCL. However, only 2.9% of the females from the fast growing, least dense populations were so protected.

January 1 of each year is designated as the "birthdate" of a marron. After hatching, young marron undergo two molts before they come to resemble their parents and become independent of the mother.

Marron are preyed upon by a number of aquatic and semi-aquatic predators. Hutchings (1981) reports that larger predaceous hemipteran and aniospteran nymphs (insects) may prey on juvenile marron. Predatory fishes are a major source predation. In eastern Australia, eels (*Anquilla* sp.) can easily gain entrance to unprotected marron ponds and consume many marron. In southwestern Australia, the introduced redfin perch (*Perca fluviatilus*) has been found to seriously limit marron populations in larger irrigation reservoirs (Morrissy 1978a). Predatory fish-eating birds, including darters (family Anhingidae) and cormorants (family Phalacrocoracidae), are considered to be amongst the worst predatory pests in Australian marron culture (Hutchings 1981) and the cost of protecting ponds from such predators can be great (Reynolds 1984). The water rat (*Hydromys chloragaster*) is also identified as a potentially serious marron predator in both eastern and southwestern Australia (Hutchings 1981; Reynolds 1984).

DISEASES

There are no serious diseases of marron either in natural populations or culture (Morrissy 1976a, 1983a). A number of diseases common to other Australian crayfishes have been noted and are covered more completely by Mills (1983). Protozoan species have been noted to occur on the body surfaces of marron and include *Lagenophrys deserti* (Kane 1964), a species of *Cothurnia*, and colonies of *Epistylis* species (Evans 1986). Although no microsporidian diseases have been noted in marron, other species of *Cherax* in Western Australia have been found to have *Thelohania* infections and, therefore, marron may be infected. The occurrence of temnocephalids has been noted by both Kane (1964) (*Temnocephala marcovichi*) and Evans (1986) (species not identified). Nematodes have also been observed on the gills of marron and appear similar to those on yabbies (Evans 1986) but as yet no identification has been made.

Hart and Hart (1967) noted the occurrence of ostracods in the gills of *Cherax* species from Western Australia, although no reference was made to the particular species on which the ostracod species occurred.

The ubiquitous species, *Psorosperium*, has been observed in marron. Cysts of this organism have been found in the muscle, gut tissue, hepatopancreas, and gills of marron (Evans 1986); however its effects, if any, are unknown.

A new species of microsporidion, *Vavria parasticida* has recently been described by Langdon (1991).

NATURAL FISHERIES

Natural fisheries for marron are restricted to the waters of Western Australia at this time as there are no natural sustainable populations elsewhere. The fishery is heavily regulated and restricted to sportsmen only. Wild marron cannot be sold legally and cultured marron must equal the legal size limit of 76 mm CL (55 mm OCL) to be sold as food in Western Australia (Morrissy 1983c; Evans 1986). The season begins in the middle of December and ends at the end of April. The bag limit is 20 per day. This size limit is not based on sound biological grounds but rather on sociological consider-ations. The general size limit was a size judged many years ago as being of a good eating size and coincidentally the same size as that of the marine rock lobster harvested commercially in Western Aus-tralia. Four legal marron, at least 120 g each, constitute one "serv-ing." The average family has four members so a day's legal catch would provide a meal for that family. However, the size limit does not take into account slow-growing populations where relatively few marron actually reach the legal size. Such populations will remain "stunted" unless exploited. Exploitation could be permitted on a population-by-population basis but could not be effectively policed with available resources to prevent capture of smaller cray-fish from "healthy" populations with the claim that they came from targeted populations.

The sports catch of marron is 100-200 tons annually (Morrissy 1978b, 1984). Lower catches correspond to years of low water. Because marron seldom forge during the day, fishing is generally done at night (Morrissy 1978b; Olszewski 1980) with baited lift nets (6 per fisherman), "snares," or hand-held scoop nets. The best marron seasons are said to be early or late summer. Variation in catchability with drop nets has been correlated mathematically with underwater illuminance (bright-negative), moon phase (dark-posi-tive), sex (male-positive), female spawning activity (gravid-nega-tive), molt stage (intermolt-positive), and previous history of cap-ture (yes-negative) (Morrissy 1970, 1973, 1983d; Morrissy and Caputi 1981). Poultry feed in a bag or pieces of fish are typical baits (Morrissy 1978b; Olszewski 1980). It is common to use bait when fishing with hand nets or snares. Snares are made using a thin, flexible wire loop supported by a straight pole. The loop is guided around the marron's abdomen toward the cephalothorax and when the loop reaches this body midpoint, it is quickly tightened and the crayfish is then lifted from the water. Drop nets are very effective in deeper waters but, as a conservation measure, they cannot now be used from a boat. Divers are no longer allowed to take marron for the same reason.

Marron may be held alive for several days out of water as long as they are damp and cool. They are not as hardy, however, as burrowing grayfish species.

AQUACULTURE

Marron aquaculture in Western Australia is severely constrained by lack of suitable water. It is further limited by the species' sensitivity to low oxygen/eutrophic waters and higher temperatures. Three levels of production are recognized: hatcheries; intensive, high density culture in small ponds (less than 1.0 ha); and extensive, low density culture in small earthen farm ponds (less than 1.0 ha). Little effort is apparently devoted to establishing sustaining populations in large water bodies as is the case in North American crayfish culture because few water bodies exist that can be used for this purpose. Whether culture can develop in areas such as the Northern Hemisphere is highly speculative based on the apparent high susceptibility of *Cherax* spp. to the crayfish fungus plague (Morrissy 1983c).

There are many references dealing with various levels of marron aquaculture (e.g., Morrissy 1976a, b, 1980, 1983d; Morrissy et al. 1990). Hatchery techniques have been fully developed and were fully described earlier by Morrissy (1975a, 1976b). Bennison (1984) discusses further modifications for recovery of newly released young of the year. The following discussion is based on these references.

Adult broodstock of similar size are held in earthen ponds at densities about $1.0/m^2$ at a male:female sex ratio of between 1:1 and 1:5. They are fed as described later for intensive pond culture. Seventy-five percent or more of the females can be expected to mate and spawn. An average female of legal size, 120 g, should produce 350 offspring. Shortly before mating begins in the spring, broodstock may be transferred to tanks (Bennison 1984) or into fenced areas within a growout pond (Morrissy 1976b). If the tank method is chosen, circular tanks (3.6 m diameter) may be stocked at a density up to 12 adults/m^2 as long as each marron has individual housing (Figure 82).

Following mating, noted by the presence of a spermatophore on the sternum, females are transferred to another tank with a 1.2 cm^2 mesh false bottom at densities up to 15/m^2. Young drop through the screen into a mass of artificial fibers or nylon substrate and are recovered by draining the tank either directly into a growout pond or a catching basket (Figure 83). This method is labor intensive and costly and subject to problems associated with mechanical breakdowns; however, each 3.6 m diameter tank can be expected to produce 50,000 young, enough to stock a 1.0 ha growout pond.

If a fenced area is to be used as a "hatchery," spawning females are transferred from a broodstock pond. Appearance of newly released marron is monitored with synthetic weed samplers. Once release is judged to be complete, the water level is dropped to recover parent females. Number of young released is estimated on the basis of the size of each female. The hatchery area should be much longer than wide to provide for maximum escapement for young marron.

Aquaria or other small containers may be used for spawning marron where low density (0.5-1.0/m^2) stocking is contemplated for extensive culture in small farm ponds. Confined females are fed sparingly (Morrissy 1970) to avoid microbial contamination of incubating eggs and to maintain the water quality. If eggs become contaminated prophylactic chemicals used to treat fish eggs may be useful if such use is permitted by health authorities. However, wherever possible, a competent fish pathologist should be consulted if a problem arises.

It is possible to advance spawning in marron with subsequent early release of young by manipulating temperature and photoperiod (Morrissy 1983a). (An indoor pool for controlled marron breeding is shown in Figure 84.) Mature males and females will mate in July (mid-winter) when held at 18-20°C on a 12:12 light-dark photoperiod. Following spawning, young can be obtained around mid-October, about two months earlier than under natural conditions. Increasing water temperature is more important than day length (summer day lengths greater than 14 hours inhibit maturation

Figure 82. Round tanks used for breeding/hatching and nursing marron. J. Huner.

of ovaries). Mating commences in ponds at temperatures around 11°C when day length is increasing and has reached 12 hours.

Intensive pond culture of marron has not yet arrived at the potential predicted by the comprehensive growout studies reported earlier by Morrissy (1979). He noted potential harvest of 2100 kg/ha, 45 g mean weight, and 3175 kg/ha, 111 g mean weight, for the first and second years of life, respectively, for marron stocking rates of about 5/m^2. More recently Morrissy (1992) reported on an intensive study of marron grown in 100m2 earthen ponds with 12 single cohorts stocked at both 5 and 10/m2 as juveniles. Growth was monitored at 4 to 5 month intervals for a period of 2 years and harvested in November-December. Marron were fed trout pellets at a rate ranging from an initial 10g/m2/week to a maximum of 30g/m2/week adjusted for marron growth. Final weights of harvested marron ranged from 47 to 156g at final densities of 0.7 to 4.5 marron/m2.

Figure 83. Marron nursery tank filled with vegetable sacking material which serves as substrate for hatchlings. J. Huner.

Growth was found to be density dependent with almost 90% of the mean body weight attained at 2 years of age being gained after the initial 300 days of growout. The first 300 days of fast growth were associated with high mortality rates whilst the second 300 days of fast growth were associated with low mortality. Harvest biomass ranged from 1100-3000 kg/ha and increased linearly with final density, i.e., high final densities gave higher production. Morrissy suggests from this study that the initial growout period of 300 days could be carried out in nursery ponds, with the second 300-day growout being carried out in growout ponds removing market-size individuals over this period to reduce densities and increase growth. It was also pointed out that growth achieved in this study was achieved at temperatures that ranged from 10-23° C and represents 36% of possible maximum growth that could be achieved at the optimum temperature of 24° C. Selection of a more suitable loca-

Figure 84. Breeding/hatching tank inside of the marron hatchery. Note the individual hides for females and mats of vegetable sacking material for the hatchlings. J. Huner.

tion than used in that study should increase growth rates and increase yearly production rates.

Farmers had difficulty achieving these production levels in practical growout units. Production from marron farming for 1989/90 was only 12tns (Treadwell et al. 1991). They have also been faced with the prohibition of the sale of marron under 120 g in size in Western Australia, a conservation measure aimed at protecting wild stocks. It would be economically preferable to harvest and sell the crop after one year and an extensive cost analysis of marron aquaculture has been provided to justify modification of existing regulations to permit the harvest and sale, for food, of 45 g, one-year-old marron (Morrissy et al. 1986). Such marron are sold for food in eastern Australia (Hutchings 1984) where no legal restrictions apply. Recently this restriction was lifted to permit sale of small marron (Morrissy et al. 1990).

Feeding marron is simple and inexpensive but creates problems for maintaining water quality. Basically, a plant detritus base is established in the pond. Early studies utilized poultry laying mash as food but this resulted in unacceptable biological oxygen demands (BOD) and resultant low dissolved oxygen levels. Lucerne in the form of pellets was found to be the most economical food that minimized BOD problems (Morrissy 1979). The amount of food depends on age; for very young marron, a feeding rate of 10 $g/m^2/wk$ for a density of $5/m^2$ gives a good growth rate over the first 3-4 months of life. When such crayfish reach 5 g in size, the rate needs to be raised to 10-25 $g/m^2/wk$. Feeding levels of 30-50 $g/m^2/wk$ are the highest sustainable rates combined with aeration during the heat of the summer when BOD is greatest (Morrissy et al. 1986). Due to the lack of basic nutritional information Morrissy (1984) tested a number of artificial diets, both commercial and experimental, for marron. The study found that molt frequently of small juveniles (0.5-5g) was less than 50% (7-44%) of the maximum molt rate of marron in growout ponds. Further it was found that with time artificial diets led to loss of appetitie and symptoms of malnutrition in marron. Morrissy (1989) tested two crustacean reference diets with similar results to the previous experiment and concluded these diets, developed for lobster, do not have the nutritional requirement for marron and require species-specific modifications to be useful. Aeration and/or flushing systems are clearly needed to ensure success of intensive marron culture (Morrissy 1979; Morrissy et al. 1984, 1986).

Optimal growth is achieved when water temperatures are in the 20-25°C range (Crook 1981). In the vicinity of Pemberton, Western Australia, marron gained 36 g during a December-April period, 15 g during the August-December period, but only 4 g during a April-August period, the coldest months of the year.

Pond construction is straightforward (Morrissy and House 1979), with recommended sizes between 0.1-1.0 ha and 1-2 m deep (Figures 85 and 86). Smaller ponds are dictated by intensive management involving stocking of known numbers of marron and the difficulty in securing suitable land and water resources in Western Australia. Rapid exchange of water may be necessary to remove excess nutrients that lead to high BODs and low dissolved oxygen. Deeper ponds are protected from excessive water temperatures in the heat of the summer. It is generally suggested that for every 1.0 ha of growout ponds, 0.1 ha of broodstock ponds need to be maintained. Elaborate fencing often with electric shock systems are used to exclude predators and prevent marron from escaping from culture ponds (Figure 87).

Extensive marron culture is typically practiced by stocking juvenile marron at low densities, $1.0/m^2$, or fewer, in small earthen farm ponds (Olszewski 1980). Pond size is usually 1530 m^3 and 0.14 ha with maximum depths of 3.7-4.3 m. Eutrophic ponds are unsuitable for marron culture as thermal stratification in the summer will result in lethal oxygen levels in deeper waters where temperatures are most favorable for marron growth and survival. Run-off carrying plant detrital and animal manures must be excluded during the summer to reduce eutrophication and associated high BOD problems. However, new or renovated ponds will often be deficient in detrital biomass and organic matter may have to be added to generate a detritus-based food chain. Cover is provided in the form of pieces of pipe or logs in deeper waters in clear ponds, however, this is not needed in turbid ponds. Cover and/or turbidity helps to minimize intra- and interspecific predation. Biomass after 1 or 2 years is usually around 300 kg/ha but 1000+ kg/ha has been recorded (Morrissy 1980). Spawning will take place when marron reach maturity,

Figure 85. Juvenile marron production pond. Note netting, paddlewheel aerator, and flashing around edge of pond to prevent crayfish from escaping. J. Huner.

which can lead to excessive numbers of young and resulting poor growth and stunting. Therefore, it is suggested that marron be completely harvested every several years. The presence of perpetuating marron populations in larger Western Australian irrigation and water storage reservoirs suggests that extensive marron culture on a large scale is possible through the establishment of sustaining populations where suitable sites are available.

Regardless of level of marron culture, marron may be harvested with hand drawn seines, baited lift nets, or traps (Morrissy 1980). Pond draining is used to ensure complete harvest of marron. Morrissy (1975b) discusses factors affecting catchability of marron. Larger, intermolt, male crayfish are most vulnerable to baited lines, nets, and traps.

Figure 86. Marron production ponds. Note netting. J. Huner.

Section VI:
Redclaw, *Cherax quadricarinatus*

TAXONOMY

The valid scientific name for redclaw (Figure 88) is *Cherax quadricarinatus* (Clark 1936) and *Cherax bicarinatus* (Riek 1969). In a recent taxonomic revision Austin (1986) considered the two species indistinguishable, but the work remains unpublished. Using only taxonomic references the synonomy of *Cherax quadricarinatus* and *Cherax bicarinatus* is as follows:

– *Astacus bicarinatus* (Gray 1845; Erichson 1846)
– *Astacus quadricarinatus* (Martens 1866)

– *Cheraps quadricarinatus* (Faxon 1898; Ortmann 1902; Calman 1911)
– *Chaeraps quadricarinatus* (Smith 1912; Faxon 1914)
– *Cherax quadricarinatus* (Clark 1936; Riek 1951)

DISTRIBUTION

The redclaw is a large species inhabiting mid- and eastern Northern Australia. Its range has recently been extended with transplants

Figure 87. Elaborate predator guards around the margin of marron ponds include an electric fence, flashing, fine mesh screen, and netting. J. Huner.

into southern Queensland and New South Wales for aquaculture purposes (P. Horwitz, unpublished data) but it appears that feral populations have not been established in local streams. This species has been translocated to overseas countries, principally the USA.

MORPHOLOGY AND ANATOMY

The redclaw is a moderately large *Cherax* species reaching in excess of 300 g and measuring 219 mm total length. The color of the exoskeleton is generally green-blue and slightly mottled but a blue color has been observed. The redclaw can be distinguished from other sympatric *Cherax* spp. by the presence of an uncalcified patch and often red colored membrane on the underside of the chelae in males as well as the absence of a median carina on the dorsal surface of the carapace. The chelae are narrow in both males

Figure 88. Large redclaw, *Cherax quadricarinatus.* R. Hutchings, FACT, Kalbar, Queensland, Australia.

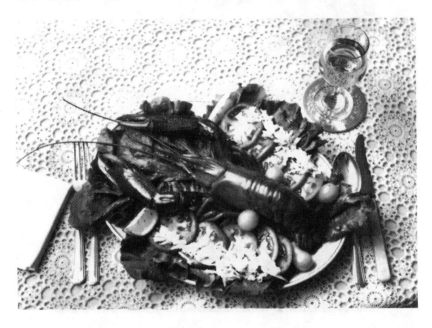

and females (Horwitz 1990). The amount of flesh in the tail of the redclaw is approximately 22% of total weight (Jones 1989).

LIFE HISTORY AND ECOLOGY

Although this species has received considerable attention in recent years for its aquaculture potential, little is known of its natural biology.

The redclaw is found in permanent water where it can occur in habitats ranging from shallow, clear, rocky, and flowing creeks, to deep, turbid clay based billabongs, and water holes on river courses. Redclaw has often been reported as nonburrowing (Hutchings 1987) but recently a low level of burrowing has been reported (Jones 1989). According to Horwitz and Richardson's (1986) burrow classification, type 1a would best describe redclaw burrows making this species similar to marron (Jones 1989). No studies made of distribution of redclaw in its natural habitat have been conducted. Under natural conditions food eaten by redclaw is mostly decaying plant and animal material and associated bacteria and fungi which cause the decomposition (Jones 1989).

From a limited analysis of a population in northern Queensland (Jones 1989) it has been suggested that redclaw will reach 70 g in 4-5 months after spawning in October/November, 110 g at 1-4 years and up to 200 g in 2-4 years, with most of the growth occurring in the warmer months. Optimal temperature for growth is 24-28°C with growth and survival significantly reduced at 34°C and growth significantly reduced at 20°C (Jones 1989). The most suitable climate for this species would seem to be subtropical.

Tolerance of redclaw to various environmental conditions has received little attention. A study of salinity tolerance has shown the redclaw is able to tolerate salinities up to 12 ppt and possibly exceeding this, while above 18 ppt the crayfish becomes lethargic and weight loss occurs (Jones 1989). Redclaw appear particularly tolerant to low levels of dissolved oxygen respiring normally at oxygen concentrations as low as 1 ppm, but for long-term health and performance, 5 ppm O_2 or above is recommended (Jones 1989).

Sexes are separate in redclaw as in other crayfish species. The development of the ovary and spawned eggs has been described (Jones 1989). Mating in this species usually takes place at night (Jones 1989) and has been recently described (Sammy 1988). The male manipulates the female on top with the ventral side together and the male deposits a spermatophore on the sternum between the walking legs. After mating, which lasts about one minute, the female spreads the sperm mass between the base of the walking legs by using the tips of the last pair of walking legs. Eggs are laid within 24 hours with brooding behaviour being similar to that described for other crayfish (Sammy 1988).

In northern Australia breeding takes place up to three times in the "wet season" at monthly intervals from October to December. After release of juveniles in February the females molt, with the next two breeding cycles beginning in mid-winter (Sammy 1988).

One unusual feature of this species is the "nonaggressive" nature of adults, although juveniles are highly cannibalistic (Sammy 1988; Jones 1989), making it an ideal aquaculture species, although this is disputed by others.

AQUACULTURE

A substantial number of persons have invested in marron culture in Queensland, although the results have generally been disappointing. Marron require better water quality than such species as yabbies, low temperatures and high oxygen levels. Recently the redclaw has attracted considerable attention among entrepreneurs (Fallu 1987/88) as a substitute species for marron. Although a commercial industry has been in operation since early 1987 there is little or no information available from these operations. Recently a comprehensive report was released by the Department of Primary Industries in Queensland (Jones 1989) and this report forms the basis of the discussion below.

The system of culture of redclaw is similar to that described for the yabbie and is of the semi-intensive type, having three stages: hatchery, nursery, and growout phases.

Hatcheries are not currently used for commercial production of juveniles for pond stocking but a system similar to that employed in marron farming is used, i.e., a broodstock pond is employed from which juveniles are captured by retrieving plastic mesh habitats (refuges) which have been placed in the growout ponds. There is some concern regarding the survival of juveniles using this system and it is thought that survival is as low as 5-10% with high degree of variable growth occurring (Jones 1989). Production of juveniles is also limited to the six month natural breeding season and is also unpredictable (Jones 1989). A hatchery and nursery can overcome many of these problems and Jones (1989) has developed simple methods for stimulating spawning, incubating eggs, and growing juveniles for stocking in growout ponds.

The hatchery building can be of any size. The hatchery developed used 2 m diameter tanks with a water depth of 40 cm, with a flow-through system allowing a water exchange rate of twice per day. The tanks have a sand substrate with lengths of PVC pipe which act as hides for the adults. The water is aerated and maintained at 25-26°C and photoperiod set at 14L:10D. Prior to placing females in the hatchery tanks their ovaries are checked *in viva* for maturity using the methods developed by Jones (1989). Using a sex ratio of 1:4 (male:female) and 14 females per tank a spawning success rate of over 90% is achieved with 80% spawning within the 10 weeks after introduction. Variability in spawning time has been noted ranging from 7 days to 100 days. Feeding of broodstock is easily achieved using water plants (e.g., *Potamageton*). Females are checked once a week to minimize disturbance to the female and eggs.

Once it is observed that the juveniles are about to leave the female, the females are transferred to nursery tanks for initial grow-out prior to stocking of growout ponds. The prime reason for trans-ferring females with attached young to the nursery stage is that newly hatched juveniles are extremely sensitive to handling. There-fore, it is best to move the females with attached young. Nursery tanks constructed of fiberglass with dimensions of 1.8 × 1.0 m are used and covered with shade cloth. Water is continuously supplied to the tanks at a rate that allows two complete water changes per day. Each tank has a substrate of fine sand which is covered by 30 cm of water. To maximize surface area, 3 cm-wide fiberglass fly-screen strips are suspended in the water in numbers to occupy 20% of the water volume.

Three to five females are stocked in each tank and are checked every few days for juvenile release after which the females are removed. To minimize release time differences, females at a similar stage of prejuvenile development are introduced to the tanks. Stocking rates of $1200/m^2$ result in 52% survival.

The best feed for newly released juveniles appears to be zooplankton consisting of cladocerans, copepods, and chironomid larvae. Feed is best presented in the frozen form as it sinks and is readily available to the crayfish. Feeding rates start at 22% of total biomass (on a wet weight basis) to 80% of biomass at harvest at mean weight of 0.1-0.4 g after an average of 42 days. One feature in common with most crayfish species is that growth is highly variable in the nursery, with up to 42% difference in growth between smallest and largest crayfish.

The growout phase is carried out using earth ponds into which post-nursery juveniles are stocked. Although 9 tons were produced from commercial operations in Queensland in 1988/89 (Treadwell et al. 1991) there were no published accounts of the methods involved, until recently when Jones (1989) gave a brief description of the methods used. Ponds used range from 500 m^2 to greater than 1 ha and are 1-2 m deep. Juveniles from brood stock ponds are stocked at densities of between $2-10/m^2$ and grown to commercial size over a period of 1-3 years. Feeding varies with intensity of the farm and includes fresh vegetation or agricultural by-products and generally a pellet ration, usually a chicken pellet type. Aeration is commonly provided using mechanical agitators, usually paddlewheels. Artificial shelters are also often provided and range from lengths of pipe to old car tires. Culling of the population is not generally practiced within the growout period, but rather the whole crop is harvested once market size is reached. Harvesting is achieved by trapping and/or by pond draining.

To prevent predators entering ponds, bird netting is often used to cover ponds and small fences are constructed around ponds and covered with electric wire to prevent predation by water rats.

There are no production figures available from farms but reports of up to 3000 kg/ha have been made (D. O'Sullivan, personal communication, 1990).

In experimental pond trials Jones (1989) found that stocking juveniles at between $5-11/m^2$ in earth ponds, 0.12 ha in size, and fed a mixture of lucerne pellets, marron pellets, and sorghum resulted in growth to 35 g with between 54-74% surviving by the end of the 250 day growout period (December-August/September). Production levels achieved ranged from 1200-1900 kg/ha. Projected production rates over one 12-month growth period ranged from 1800-2600 kg/ha at 60 g mean weight. From the experimental study it was concluded that production in excess of 2000 kg/ha is achievable with stocking rate 5/m² using nursery produced juveniles. Experimentation with various feed types for growout found that lucerne pellets offer a good diet base as the feed rapidly decomposes to form a detritus base for the crayfish. Studies by Jones (1989) using a variety of plant, animal, pellet, and pet feeds found superior growth for the formulated diets although the economics of feeding formulated diets in a commercial situation is still in question until a specific formulated feed can be developed for redclaw.

Section VII:
Economics

Although technical investigations of Australian freshwater crayfish for their aquaculture potential commenced in the 1960s, it was not until the late 1980s that any significant economic analysis of crayfish farming was made. Since that time three economic studies have been published examining yabbie, marron, and redclaw farming but each is largely based on hypothetical information due to lack of any significant commercial farming operations. Outlined below is a brief summary of each of these studies.

Standiford et al. (1987) describe a yabbie culture farming operation scenario that could generate an internal rate of return of 23.7% and the rate of return of capital invested of 27.4% assuming a price of $10.00 (Australian) per kg. This study calls for 10 ha of intensive rearing ponds (0.2 ha each, 50 total). One million juveniles would be required annually necessitating hatchery production of two million young (50% survival in nursery ponds). The general management program, hatchery, nursery ponds, and growout ponds, has been described above. Harvest size is approximately 55 g. Total production would be 33,000 kg per year. A processing plant is included for cooking and freezing yabbies to be packed in 2 kg units

Table 23. Capital costs and annual budget of establishing a hypothetical 10ha yabbie farm (1987 values) (modified from Staniford et al. 1987)

CAPITAL COSTS	
ITEM	$A
Land	47,000
Pond establishment	173,000
Water storage dam	40,000
Hatchery and office	27,535
Nursery	64,625
Processing plant shed	19,000
Storage shed and equipment	3,500
Predator control	53,200
Electricity connection	10,000
Monitoring and laboratory equipment	4,000
Pond transfer equipment	4,000
Feeders	2,000
Pumps	800
Harvesting and processing equipment	8,000
Transport equipment	23,000
Total Capital	**479,660**
ANNUAL BUDGET	
Item	$A
Gross Income: 33,000 kg @ $10 per kg	330,000
Costs	
Feed	31,050
Hatchery water	306
Energy	10,000
Packaging	8,250
Repairs and maintenance	
Equipment	8,080
Buildings	275
Labor	96,000
Insurance	4,797
Vehicle registration and insurance	1,000
Rates and taxes	1,500
Accounting fees	500
Depreciation	33,612
Office costs	1,300
Sundry expenses	2,000
Total Costs	**198,670**

in cardboard cartons for retail outlets. An annual budget and capital costs for a hypothetical yabbie farm is shown in Table 23.

Morrissy et al. (1986) describes a marron culture scenario very similar to that described by Standiford et al. (1987). Size at harvest would be in the 40-50 g size range with similar stocking rates to yabbies for a facility with 10 ha of culture ponds averaging 0.2 ha each. This would generate a return on capital of 10-20%. Marron grow significantly larger than yabbies, but it takes over 18 months to reach sizes in excess of 100 g. It is apparently unprofitable to grow them to larger sizes. Annual budget and capital costs are shown in Table 24.

An economic analysis for redclaw culture is described by Cann and Shelley (1990) based on 6 ha of ponds which will generate an internal rate of return of 16.7% with a farm gate price of $12/kg. Newly hatched redclaw from a hatchery are stocked in 0.025 ha juvenile ponds with a 70% mortality rate and subsequently stocked in 0.5 ha growout ponds with 50% survival. Production from the growout ponds is estimated to be 3000 kg/ha. Capital costs and annual budget are given in Table 25.

With the rapid development of crayfish farming in the late 1980s and early 1990s there has become available a considerable amount of detailed information on the economics of this form of aquaculture. A recent study by the Australian Bureau of Agricultural and Resource Economics (Treadwell et al. 1991) has gathered this information based on industry input to provide a comprehensive report on the current status of crayfish farming in Australia, a summary of which is given below.

The authors examined marron, yabbies, and redclaw with a farm size of 15 ha of growout ponds which allowed direct comparison between the species (Table 26). The major difference between the species were stocking and growth rates. Stocking rates used range between 5-10/m^2 with yabbies stocked at the lowest level, due to their aggressive nature, and redclaw at the highest level. Redclaw are marketed at 80 g after approximately 8 months of growout, marron take 12 months to achieve market size of 70 g, and yabbies are marketed after 9 months at a small size of 60 g.

The model used for marron farming is based on a hatchery to produce juveniles for stocking growout ponds in the second year of

Table 24. Capital costs and annual budget for establishing a hypothetical marron farm (1986 values) (modified from Morrissy et al. 1986)

CAPITAL COSTS	
ITEM	$A
Land (20 ha – Margaret River, Western Australia)	30,000
Pond establishment	116,000
Water and drainage system	118,000
Manager's home (transportable)	34,200
Garage and storage shed	4,300
Processing shed and equipment	32,000
Monitoring equipment	1,300
Harvesting, feeding, and aeration equipment	117,000
Water supply	107,000
Predator control	150,000
Total Capital	709,900
ANNUAL BUDGET	
Item	$A
Staff (manager, assistant, casual labor)	54,200
Marketing (telephone, advertising, packaging)	7,700
Insurance	1,700
Registration and licences	400
Electricity	3,400
Fuel, etc.	5,100
Total Costs	72,500

Note: Additional annual costs unaccounted for in this table are: interest and loan repayments, depreciation on capital items ($3,000-$3,500/ha/yr), costs of repairs and maintenance, land rates. Costs of feed at $0.30/kg and seed costs at $0.10/marron are also not included in the table.

production. In the first year of operation it is assumed that 75,000 juveniles are purchased for 70 cents each for pond stocking, and breeding stock for the hatchery are bought for $10 each. The model for redclaw is similar to marron but due to the early stage of development for this species low production is projected for the first two years, with full production in year three. In the first year 600,000 juveniles are purchased at 35 cents each to stock half the ponds with an initial production of 1.5 ton/ha. Mature broodstock are also

Table 25. Capital costs and annual budget for establishing a hypothetical 6ha redclaw farm (1990 values) (modified from Cann and Shelley 1990)

CAPITAL COSTS	
ITEM	$A
Land	40,000
Access road	2,000
Electricity connection	20,000
Building	25,000
Single quarters	12,000
Earthworks	118,000
Well, pump, and motor	20,000
Piping	13,426
Fencing	9,110
Netting	17,384
Aerators	35,000
Electricity connection to aerators	20,000
Plant and equipment	47,750
Total Capital	**391,790**

ANNUAL BUDGET	
Item	$A
Variable Costs	
Pond preparation	6,656
Feeding	10,312
Pumping	14,157
Aeration	26,312
Repairs and maintenance	10,781
Overhead Costs	
Labor and on-costs	60,260
Administration and sundry costs	10,000
Total Costs	**138,478**

purchased in the first year for hatchery production. The yabbie farming model used differs from the other species in that no hatchery is specified, rather juveniles are harvested from growout ponds, where natural reproduction has occurred, and restocked in other ponds for growout. (Note: this is not the usual method used by farmers; usually a hatchery phase is employed [see yabbie section].)

The farm is developed over a two-year period with 560,000 juveniles purchased for 5 cents each in the first year. After harvesting sufficient mature stock is left in ponds for breeding to produce juveniles for subsequent stocking.

Capital and operating costs for farming each species are given in Tables 27 and 28. Calculation of internal rate of return revealed a variation between the species (Table 29) with redclaw having better prospects compared with marron or yabbies. The external rate of return for each of the species varied with changes in production, price, and labor costs.

Treadwell et al. (1991) conclude that because crayfish farming in Australia is still a relatively new industry, technology can be further refined to produce a more viable farming operation, in particular reduction in labor costs and improved expertise to run larger farms. They point out, however, viability is sensitive to changes in price and yield.

Although yabbie, marron, and redclaw farming appear to be an attractive investment, there are several factors that can markedlyinfluence returns (Morrissy et al. 1986; Standiford and et al. 1987; Standiford and Kuznecous 1988). Some of these factors are sensitivity of profitability to production level and price, the possibility pro-

Table 26. Key characteristics of the Australian freshwater crayfish farm models (from Treadwell et al. 1991)

Item	Unit	Yabbie	Marron	Redclaw
Total size of farm	ha	23	25	24
Number of ponds	no.	75	80	77
Pond size	ha	0.2	0.2	0.2
Total area of growout ponds	ha	15	15	15
Time for growout	months	9	12	8
Harvest size (range)	g	60(50-90)	70(60-100)	80(60-100)
Yield (range)	t/ha	1.5(1-2.5)	2(1.5-2.5)	3(2.5-4)
Total farm production (range)	t/y	22.5(15-37.5)	30(22.5-37.5)	45(37-60)
Prices—domestic (range)	$/kg	12(10-16)	14(12-16)	14(12-16)
Prices—export (range)	$/kg	17(14-21)	17(14-21)	17(14-21)

Table 27. Annual operating costs for Australian freshwater crayfish farm models (from Treadwell et al. 1991)

Item	Yabbie $A,000	Marron $A,000	Redclaw $A,000
Feed	13.5[a]	26.0	26.0
Fertilizer, hay, and lime	0	0	15.0
Electricity and fuel	18.0	30.0	30.0
Labor			
owner/manager	30.0	30.0	30.0
permanent	46.0	78.5	78.5
casual	12.5	10.0	31.3
Packaging	22.5	30.0	45.0
Freight[b]	47.3	63.0	94.5
Marketing	26.2	34.9	52.3
Repairs and maintenance	22.0	28.0	28.0
Administration	6.0	6.0	6.0
Miscellaneous	5.0	5.0	5.0
Total	249.0	341.4	441.6

a. Feed cost is lower for yabbie due to the higher reliance on natural feed with the lower stocking densities.
b. based on 50% of production being exported and freight costs of 60-80c/kg for domestic sales and $3-4/kg for exports.

duction levels will vary between years, and the lack of practical management skills by fish farmers in Australia. In addition unforeseen problems such as human error, mechanical breakdown, and production failure can seriously affect profitability (Kingsley 1986).

Section VIII: Marketing

Almost all yabbies produced from aquaculture are supplied to the domestic market and mainly directly to restaurants, although a small proportion is also sold as bait. At present, production of this species from aquaculture is greatest for all *Cherax* spp. with approximately

Table 28. Capital costs for the Australian freshwater crayfish models (from Treadwell et al. 1991)

Item	Yabbie $A,000	Marron $A,000	Redclaw $A,000	Scrap value %	Life y
Land	69.0	75.0	72.0	100	—
Perimeter fence	11.1	14.3	13.8	5	20
Office and storage shed	25.0	25.0	30.0	50	20
Electricity connection	20.0	20.0	20.0	100	—
Generator	8.0	8.0	8.0	10	10
Water storage dam	30.0	30.0	30.0	—	20
Ponds, channels, gates, screens[a]	225.0	240.0	231.0	—	20
Pumps and motors	25.0	25.0	25.0	10	5
Piping	75.0	80.0	77.0	—	10
Netting and shelter in ponds[a]	26.3	28.0	27.0	—	3
Posts and wires for netting	30.0	32.0	30.8	—	20
Fences in each post	0	0	61.6	10	10
Aerators	16.5	84.7	87.0	10	5
Truck/utility	25.0	25.0	25.0	10	10
Tractor	10.0	10.0	10.0	10	10
Motorcycle	5.0	5.0	5.0	10	5
Fertilizer spreader	0	0	2.5	10	5
Slasher	1.5	1.5	1.5	10	10
Bucket	1.0	1.0	1.0	10	10
Blade	1.0	1.0	1.0	—	10
Feed blower	1.0	1.0	1.0	—	3
Testing equipment	2.0	2.0	2.0	—	3
Harvesting equipment[b]	2.5	2.5	2.5	—	3
Holding tanks and shelter	6.0	12.0	15.0	10	10
Coolroom/freezer[b]	10.0	10.0	10.0	10	10
Processing room[b]	20.0	20.0	20.0	—	20
Processing equipment[b]	5.0	5.0	5.0	—	5
Office equipment	5.0	5.0	5.0	10	5
Miscellaneous	5.0	5.0	5.0	10	5
Juveniles	28.0	150.0	210.0	—	—
Initial breeding stock	10.0	48.0	0	—	—
Total	698.9	966.0	1034.7	—	—

a. In addition, extra labor has been allocated to operating costs in the initial years for some construction work.
b. First purchases in year 1 for first harvest

Table 29. Internal rates of return from Australian freshwater crayfish farm models (from Treadwell et al. 1991)

Item	Mean IRR %	IRR range[a] %
Yabbie		
Current, constant prices (yield 1.5 t/ha)	9.3	2.1-17.0
– labor cost down 10%	10.4	3.5-17.8
– prices fall 1.5%/y[b]	2.6	–7.2-11.6
– yield 2 t/ha, constant prices	23.7	17.6-30.7
– yield 2 t/ha, prices decline 1.5%/y[b]	19.1	12.0-26.6
Marron		
Current, constant prices (yield 2 t/ha)	10.3	6.7-14.0
– price $18/kg, constant	18.8	14.6-23.4
– price $18/kg, declining 1.5%/y[b]	14.5	9.8-19.2
– labor cost down 10%	11.2	7.5-14.8
– prices fall 1.5%/y[b]	4.6	0.0-8.8
– yield 2.5 t/ha, constant prices	20.3	16.9-23.7
– yield 2.5 t/ha, prices decline 1.5%/y[b]	16.2	12.4-19.8
Redclaw		
Current, constant prices (yield 3 t/ha)	19.4	15.7-23.0
– labor cost down 10%	20.5	17.0-24.2
– prices fall 1.5%/y[b]	11.9	6.6-17.0
– yield 4 t/ha, constant prices	27.2	24.4-30.2
– yield 4 t/ha, prices decline 1.5%/y[b]	21.7	18.2-24.0
– yield 4 t/ha, prices decline 3.0%/y	14.0	9.0-18.3

a. Since there is a 25% chance of the IRR being below or above the lower and higher figure, there thus is a 50% chance of the IRR being within the range shown
b. Prices deflated by 1.5% a year relative to costs

97 tons produced in 1990/91 with prices obtained ranging from $Aus.8-12/kg (Table 30). Most supplies of yabbies to the market come from natural fisheries or from farm dams, both of which however are renown for fluctuating catches. The current rapid up-surge in interest in yabbie, pond acreage suggests that production in the future should increase significantly.

All legal size marron produced from aquaculture are destined for the restaurant trade. Production is currently at a low level, in

Table 30. Recent and projected production of freshwater crayfish in Australia (from Treadwell et al. 1991)

	1988/89		1989/90[a]		1990/91[b]	
	tons	$A,000	tons	$A,000	tons	$A,000
Yabbie	26	357	55	820	97	1253
Marron	2	52	12	298	10	226
Redclaw	9	200	31	496	46	690
Total	37	609	98	1513	153	2169

a preliminary figures　　　*b* forecast

1990/91 being approximately 10 tons and prices ranging from $Aus.10-30/kg with larger sizes receiving a premium price (Table 30). In addition to sale of table size marron, a considerable trade existed until recent years in producing and selling juvenile marron to other marron growers within Australia and overseas.

Since redclaw is a new species to aquaculture, production is at a low level, reaching 46 tons in 1990/91, with prices obtained being close to $Aus.25/kg (Table 30). With current interest in this species and increased pond acreage, production can be expected to dramatically increase in the near future.

There is little information on the prices that these three species can be expected to command on the European markets although unsubstantiated reports suggest prices would be high. When cooked both marron and redclaw have a red color, the preferred market color, whereas yabbies cook to an orange color which may be less preferred. Until production of these crayfish reaches levels where continuity in export can be guaranteed the price will remain unstable.

Section IX:
Status of Australian Crayfish Culture
1989-90 Season

O'Sullivan (1991) summarized the overall status of Australian crayfish culture. Total production of food-sized crayfish was 112.75

ing processes. *Australian National Parks and Wildlife Service Report Series, No. 14*, Australian National Parks and Wildlife Service, Canberra, Australia.

Horwitz, P. H. J., and A. M. M. Richardson. 1986. An ecological classification of the burrows of Australian freshwater crayfish. *Australian Journal Marine Freshwater Research* 37:237-242.

Hume, D. J., A. R. Fletcher, and A. K. Morison. 1983. *Carp program, final report. Report No. 10*, Arthur Rylah Institute for Environmental research, Fisheries and Wildlife Division, Ministry for Conservation, Victoria, Australia.

Hutchings, R. 1981. Predators. *Freshwater Australian Crayfish Traders Newsletter* (Brisbane, Queensland, Australia) 1:5-8.

Hutchings, R. (Ed.). 1984. *Proceedings of the Conference on Aquaculture* in Australia, Brisbane, 1984, Freshwater Australian Crayfish Traders, Brisbane, Queensland, Australia.

Hutchings, R. 1987. Aquaculture of *Cherax quadricarinatus*. Paper given at the Annual General Meeting of the Marron Growers Association of Western Australia, 30 November 1987.

Johnson, H. T. 1978. Fishery potential of yabbie stocks in New South Wales. Internal Report, New South Wales State Fisheries, New South Wales, Australia.

Johnson, H. T. 1979. Reproduction, development and breeding activity in the freshwater crayfish, *Cherax destructor* Clark. MSc thesis, School of Biological Sciences, University of Sydney, New South Wales, Australia.

Johnson, S. K. 1977. *Crayfish and freshwater shrimp diseases*, Texas A&M University, Sea Grant College Program, Texas Agricultural Extension Service, College Station, Texas, USA.

Jones, C. M. 1989. *The biology and aquaculture potential of Cherax quadricarinatus*. Final report submitted by the Queensland Department of Primary Industries to the Reserve Bank of Australia Rural Credits Development Project No. QDPI/8860.

Kane, J. R. 1964. The Australian freshwater Malacostraca and their epizoic fauna. MSc thesis, Department of Zoology, University of Melbourne, Victoria, Australia.

Kefous, K. 1981a. Identification and analysis of lacustrine and riverine fauna from inland sites in southeastern Australia. Part 1. Crustacea. *The Artefact* 6:35-51.

Kefous, K. 1981b. 30,000 years of yabby fishing, an archaeological record of the activities of the earliest human ancestors known in Australia. Paper given at the Third School of the Australian Freshwater Crayfish, Hawkesbury Agricultural College, New South Wales, Australia.

Kingsley, R. C. S. 1986. Aquaculture–Understanding the risk factors. *Infofish Marketing Digest* 6/86:17-18.

Lake, P. S., and A. Sokol. 1986. Ecology of the yabby *Cherax destructor* Clark (Crustacea: Decapoda: Parastacidae) and its potential as a sentinel animal for mercury and lead pollution. *Australian Water Resources Council Technical Paper*, No. 87.

Langdon, J. S., 1991. Description of *Vavraia parastacida* sp. nov. (Microspora:

tons and pond area was over 300 ha with 405 licenses issued by the authorities. Juvenile production was in excess of 6,700,000.

Marron accounted for about 11% of the total harvested biomass of food-sized crayfish. Percentages for yabbie and redclaw were 70 and 19, respectively. Production of 1 ton/ha of any of the three cultured species is readily attainable. Therefore, production is sure to increase if the farms in operation persist.

Juvenile production was not broken down by species but it was clear that the greatest number of juveniles were yabbies followed by redclaw and marron. Much of the reported production of juveniles was used on the farm where they were produced. Most were produced in nursery ponds rather than in hatcheries. Table 29 shows recent and projected production of *Cherax* spp. in Australia.

LITERATURE CITED

Anonymous. 1968. River Murray reach fishery. *Australian Fisheries Newsletter* 27(11):30-32.

Anonymous. 1973. Australian yabbies popular in Scandinavia. *Australian Fisheries* 32:6.

Anonymous. 1979. Study on diet of the yabbie. *SAFIC* 3:13-15.

Anonymous. 1986. Climate and physical geography of Australia. In *Year Book Australia Number 70*, Australian Bureau of Statistics, pp. 9-37.

Anonymous. 1990. Humble yabbie: a gourmet export. *Austasia Aquaculture* 5(1): 23-24.

Archey, G. 1915. The freshwater crayfishes of New Zealand. *Transactions and Proceedings New Zealand Institute* 47:295-315.

Austin, C. M. 1986. Electrophoretic and morphological systematic studies of the genus *Cherax* (Decapoda: Parastacidae) in Australia. PhD thesis, Zoology Department, University of Western Australia, Western Australia, Australia.

Avault, J. W., Jr., and J. V. Huner. 1985. Crawfish culture in the United States. In *Crustacean and Mollusk Aquaculture in the United States*, Huner, J.V. and Brown, E.E., Eds., AVI Publishing Co., Westport, Connecticut USA, pp. 1-61.

Barley, R. J. 1983. A comparison of the responses to hypoxia of the yabbie *Cherax destructor* Clark and the Murray crayfish *Euastacus armatus* (von Martens). BSc (Honours) thesis, Department of Zoology, University of Adelaide, South Australia, Australia.

Bennison, S. 1984. Hatchery techniques In Marron farming. *Proceedings of a workshop held by the Marron Growers Association of Western Australia (Inc.)*, October 1984, Bennison, S., Ed., Perth, Western Australia, Australia, pp. 10–12.

Cadwallader, P. L., and G. N. Backhouse. 1983. *A guide to the freshwater fish of Victoria*. Victorian Government Printing Office, Melbourne, Australia.

Calman, W. T. 1911. Note on a crayfish from New Guinea. *Ann. Magazine Natural History* 8(8):366-368.

Cann, B., and C. Shelley. 1990. Bio-economics of redclaw. In *Proceedings of a seminar: Farming the redclaw freshwater crayfish*, Shelley, C.C. and Pearce, M.C., Eds., Northern Territory Department of Primary Industry and Fisheries, Darwin, Australia, pp. 34-37.

Carrick, R. 1959. The food and feeding habits of the strawnecked ibis, *Threskiornis spinicollis* (Jameson) and the white ibis, *I. molucca* (Cuvier) in Australia. *Australian Wildlife Research* 4:69-92.

Carroll, P. N. 1981. Aquaculturists' enthusiasm for yabbies highlights potential beyond the problems. *Australian Fisheries* 40(6):23-31.

Chessman, B.C. 1983. Observations on the diet of the broad-shelled turtle, *Chelodina expensa* Gray (Testudines: Chelidae. *Australian Wildlife Research)* 10:169-172.

Chessman, B.C. 1984. Food of the short-necked turtle *Chelodina longicollis* Shaw (Testudinus: Chelidae) in the Murray Valley, Victoria and New South Wales. *Australian Wildlife Research* 11:573-578.

Clark, E. 1936. The freshwater and land crayfishes of Australia. *Memoirs of the National Museum Victoria* 10:5-58.

Clark, E., and F. M. Burnet 1942. The application of serological methods to the study of the Crustacea. *Australian Journal of Experimental Biology Medical Science* 20:89-95.

Croft, J. D., and L. J. Horne 1978. The stomach contents of foxes, *Vulpes vulpes*, collected in New South Wales. *Australian Wildlife Research* 5: 85-92.

Crook, G. 1981. *Marron and marron farming*. Extension and Publicity Section, Western Australian Department of Fisheries and Wildlife, Perth, Australia.

Davis, T. L. O. 1977. Food habits of the freshwater catfish, *Tandanus tandanus* Mitchell, in the Gwydir River, Australia, and effects associated with impoundment of this river by the Copeton Dam. *Australian Journal Marine Freshwater Research* 28:455-466.

de Kretser, D. E. 1979. Aspects of the population ecology of yabbie (*Cherax destructor*) in two Victorian farm dams. BSc (Honours) thesis, Zoology Department, Monash University, Victoria, Australia.

Erichson, W. F. 1846. Uebersicht de Arten der Gattung *Astacus*. *Arch. Naturgesch* 12:86-103, 375-377.

Evans, L. H. 1986. *Preliminary studies of parasites and commensals in decapod crustaceans in Western Australia*. Report from Curtin University of Technology, Western Australia, Australia.

Eyre, E. J. 1845. *Journal of expedition of discovery into central Australia*. 2 volumes, T. and W. Boone, London.

Fallu, R. 1987/88. Crayfish farming in Australia. *International Association of Astacology Newsletter* 10(2):6-7.

Faragher, R. A. 1983. Role of the crayfish *Cherax destructor* Clark as food for trout in Lake Eucumbene, New South Wales. *Australian Journal Marine Freshwater Research* 33:407-417.

Faxon, W. 1898. Observations of the Astacidae in the United States Museum and in the Museum of Comparative Zoology, with descriptions of new species. *Proceedings United States National Museum* 22:643-694.

Faxon, W. 1914. Notes on the crayfishes in the United States National Museum and the Museum of Comparative Zoology, with description of new species and subspecies, to which is appended a catalogue of the known species and subspecies. *Memoirs Museum Comparative Zoology, Harvard* 40:351-427.

Fleay, D. 1964. The rat that mastered the waterways. *Wildlife in Australia* 1(4): 3-7.

Forteath, N. 1985. Studies on the Tasmanian freshwater crayfish *Astacopsis gouldi*. *Tasmanian Inland Fisheries Commission Newsletter* 14(3).

Francois, D. D. 1960. Freshwater crayfishes. *Australian Museum Magazine* 13:217-222.

Frost, J. V. 1975. Australian crayfish. *Freshwater Crayfish* 2:87-96.

Geddes, M. C. 1984. Limnology of Lake Alexandrina, River Murray, South Australia and the effects of nutrients and light on phytoplankton. *Australian Journal Marine Freshwater Research* 35:399-415.

Geddes, M. C., M. Smallridge, and S. C. Clarke. 1993. Potential for a nursery phase in the commercial production of the yabbie *Cherax destructor*. *Freshwater Crayfish* 8.

Grant, E. M. 1978. *Guide to fishes*. Queensland Government, Queensland, Australia.

Gray, J. 1845. New species of genus *Astacus*. In *Eyres Journal of expedition of discovery into central Australia*, Vol. 1 (appendix), T. and W. Boone, London, pp. 407-411.

Hale, H. M. 1925. Observations of the yabbie (*Parachaeraps bicarinatus*). *Australian Museum Magazine* 2:217-274.

Hale, H. M. 1927. The crustaceans of South Australia. Part 1. *Handbook Fauna Flora of South Australia*, Government Printer, Adelaide, Australia.

Hart, C.W., Jr., and D. G. Hart 1967. The entocytherid ostracods of Australia. *Proceedings of the Academy of Natural Sciences of Philadelphia* 119(1):1-51.

Haswell, W. A. 1888. On *Temnocephala*, an aberrant monogenetic, trematode. *Quarterly Journal Microscopic Science* 28:279-302.

Haswell, W. A. 1893. *A monograph of the Temnocephaleae*. Macleay Memorial Volume, Linnaean Society of New South Wales, Australia, pp. 93-152.

Hobbs, H. H., Jr. 1974. Synopsis of the families and genera of crayfish (Crustacea, Decapoda). *Smithsonian Contributions to Zoology* 164:1-32.

Holthuis, L. B. 1949. *Zoological results of the Dutch New Guinea expedition 1939. Number 3: Decapoda Macrura with a revision of the New Guinea Parastacidae*. Nova Guinea 5 (new series):289-328.

Holthuis, L. B. 1950. *Results of the Archbold expeditions No 63. The Crustacea Decapoda Macrura collected by the Archbold New Guinea expeditions*. American Museum Novitiates, No. 1461:1-17.

Horwitz, P. H. J. 1990. The conservation status of Australian freshwater crustacea: with a provisional list of threatened species, habitats and potentially threaten-

Pleistophoridae) from marron *Cherax tenuimanus* (Smith), (Decapoda: parastacidae). *Journal of Fish Diseases* 14:619-629.

Lewis, R. B. 1976. Aspects of the life history, growth, respiration and feeding efficiencies of the yabby, *Cherax destructor* Clark with regard to potential for aquaculture. BSc (Honours) thesis, Department of Zoology, University of Adelaide, South Australia, Australia.

McCoy, F. 1888. *Prodomus of the Zoology of Victoria* Volume 2, p. 225.

McCulloch, A. R. 1914. Revisions of the freshwater crayfishes of south western Australia. *Records Western Australian Museum* 1:228-235.

McNally, J. 1957. The feeding habits of cormorants in Victoria. *Victorian Fisheries and Wildlife Department of Fisheries Contributions* No. 6.

Maher, M. 1984. Benthic studies of waterfowl breeding habitat in southwestern New South Wales. 1. The fauna. *Australian Journal Marine Freshwater Research* 35:85-96.

Martens, E. Von. 1866. On a new species of *Astacus*. *Annals Magazine of Natural History* 17(3):359-360.

Miller, B. 1976. Environmental control of gonadal cycles and egg laying periodically in Little Pied and Little Black Cormorants in inland NSW. PhD thesis, School of Biological Sciences, University of Sydney, New South Wales, Australia.

Miller, B. 1979. Ecology of the Little Black Cormorant, *Phalacrocorax sulcirostris* and the Little Pied Cormorant, *P. melanoleucos,* in inland New South Wales, 1. Food and feeding habits. *Australian Wildlife Research* 6:79-95.

Mills, B. J. 1980. Effects of stocking and feeding rates on growth, mortality and production in yabbies. *SAFIC* 4:24-25.

Mills, B. J. 1983. A review of diseases of freshwater crayfish, with particular reference to the yabbie *Cherax destructor. Research Paper* Department of Fisheries, South Australia No. 9.

Mills, B. J. 1986. Aquaculture of yabbies. In *Proceedings of the First Australian Freshwater Aquaculture Workshop*, Government Printer, Sydney, Australia pp. 89-98.

Mills, B. J., and M. C. Geddes 1980. Salinity tolerance and osmoregulation of the Australian freshwater crayfish *Cherax destructor* Clark (Decapoda: Parastacidae). *Australian Journal Marine Freshwater Research* 31:667-676.

Mills, B. J., and M. C. Geddes. 1992. Effects of salinity on growth in *Cherax destructor. Freshwater Crayfish* 8 (In Press).

Mills, B. J., and P. I. McCloud. 1983. Effects of stocking and feeding rates on experimental pond production of the crayfish *Cherax destructor* Clark (Decapoda: Parastacidae). *Aquaculture* 34:51-72.

Mitchell, B. D., and R. Collins. 1989. Development of field-scale intensive culture techniques for the commercial production of the yabbie (*Cherax destructor/albidus*). Centre for Aquatic Science, Warrnambool Institute of Advanced Education, *Report* p. 253.

Momot, W. T. 1984. Crayfish production: a reflection of community energetics. *Journal of Crustacean Biology* 4:35-54.

Morrissy, N. M. 1967. The ecology of trout in South Australia. PhD thesis, Department of Zoology, University of Adelaide, South Australia, Australia.

Morrissy, N. M. 1970. Spawning of marron, *Cherax tenuimanus* (Smith) (Decapoda: Parastacidae) in Western Australia. *Fisheries Research Bulletin Western Australia* 10:1-23.

Morrissy, N. M. 1973. Normal (gaussian) response of juvenile marron *Cherax tenuimanus* (Smith) (Decapoda: Parastacidae) to capture by baited sampling units. *Australina Journal Marine Freshwater Research* 24:183-195.

Morrissy, N. M. 1974. The ecology of marron *Cherax tenuimanus* (Smith) introduced into some farm dams near Boscabel in the Great Southern area of the Wheatbelt Region of Western Australia. *Fisheries Research Bulletin Western Australia* 12:1-55.

Morrissy, N. M. 1975a. Spawning variation and its relationship to growth rate and density in marron, *Cherax tenuimanus* (Smith). *Fisheries Research Bulletin Western Australia* 16:1-32.

Morrissy, N. M. 1975b. The influence of sampling intensity on the catchability of marron *Cherax tenuimanus* (Smith) (Decapoda: Parastacidae). *Australian Journal Marine Freshwater Research* 26:47-73.

Morrissy, N. M. 1976a. Aquaculture of marron *Cherax tenuimanus* (Smith). Part 1. Site selection and the potential of marron for aquaculture. *Fisheries Research Bulletin Western Australia* 17:1-27.

Morrissy, N. M. 1976b. Aquaculture of marron, *Cherax tenuimanus* (Smith). Part 2. Breeding and early rearing. *Fisheries Research Bulletin Western Australia* No. 17.

Morrissy, N. M. 1978a. The past and present distribution of marron, *Cherax tenuimanus* (Smith), in Western Australia. *Fisheries Research Bulletin Western Australia* 22:1-38.

Morrissy, N. M. 1978b. The amateur marron fishery in Western Australia. *Fisheries Research Bulletin Western Australia* 21:1-44.

Morrissy, N. M. 1979. Experimental pond production of marron, *Cherax tenuimanus* (Smith) (Decapoda: Parastacidae). *Aquaculture* 16:319-344.

Morrissy, N. M. 1980. Production of marron in Western Australian Wheatbelt farm dams. *Fisheries Research Bulletin Western Australia* 24:1-79.

Morrissy, N. M. 1983a. Freshwater crayfish–Parastacology in relation to fisheries, agriculture and conservation. In *Proceedings of the Eighteenth Assembly of Australian Freshwater Fishermen*, Lake Eucumbene, Australia pp. 12-21.

Morrissy, N. M. 1983b. Induced early spawning of marron. *Western Australian Marron Growers Bulletin*, 5(2):1-4.

Morrissy, N. M. 1983c. Crayfish research and industry activities in Australia, New Guinea and New Zealand. *Freshwater Crayfish* 5:534-544.

Morrissy, N. M. 1983d. Marron aquaculture. In *Proceedings First Australian Freshwater Aquaculture Workshop*, New South Wales Government Printer, Sydney, Australia pp. 81-88.

Morrissy, N. M. 1984. Assessment of artificial feed for battery culture of a freshwater crayfish, marron (*Cherax tenuimanus*) (Decapoda: Parastacidae). *Department Fisheries Wildlife Western Australia Report* 63:1-43.

Morrissy, N. M. 1986. Bibliography for *Cherax tenuimanus*, marron. *Western Australian Marron Growers Bulletin* 8(1):13-16.

Morrissy, N. M., 1989. A standard reference diet for crustacean nutrition research IV. Growth of freshwater crayfish *Cherax tenuimanus. Journal of the World Aquaculture Society* 20(3):114-117.

Morrissy, N. M. 1990. Optimum and favourable temperatures for growth of *Cherax tenuimanus* (Smith 1912) (Decapoda:Parastacidae). *Australian Journal Marine Freshwater Research* 41:735-746.

Morrissy, N. M., 1992. Density-dependent pond growout of single year-class cohorts of a freshwater crayfish *Cherax tenuimanus* (Smith), to two years of age. *Journal of World Aquaculture Society* (in press).

Morrissy, N. M., and N. Caputi. 1981. Use of catchability equations for population estimation of marron, *Cherax tenuimanus* (Smith) (Parastacidae). *Australian Journal Marine Freshwater Research* 32:213-225.

Morrissy, N. M., N. Caputi, and R. R. House. 1984. Tolerance of marron (*Cherax tenuimanus*) (Decapoda: Parastacidae) to hypoxia. *Aquaculture* 41:61-74.

Morrissy, N. M., L. E. Evans, and J. V. Huner. 1990. Australian freshwater crayfish: aquaculture species. *World Aquaculture* 21(2):113-122.

Morrissy, N. M., N. Hall, and N. Caputi. 1986. A bioeconomic model for semi-intensive grow-out of marron (*Cherax tenuimanus*). *Fisheries Management Discussion Paper*, No. 2, Western Australian Fisheries Department, Perth, Western Australia, Australia.

Morrissy, N. M., and R. R. House. 1979. Economic feasibility of intensive outdoor pond culture of freshwater crayfish in Australia. *Department of Fisheries and Wildlife Report*, Perth, Western Australia, Australia.

Newsome, A. E., P. C. Catling, and L. K. Corbett. 1983. The feeding ecology of the dingo. II. Dietary and numerical relationship with fluctuating prey populations in south-eastern Australia. *Australian Journal of Ecology* 8:345-366.

Olszewski, P. 1980. *A salute to the humble yabby*. Angus and Robertson, Sydney, Australia.

Ortmann, A. E. 1891. Die Decapoden-Krebse des Strasburger Museums, Parastacidae. *Zoologische Jahrbucher, Abteilung feur Systematik* 6:7-9.

Ortmann, A. E. 1902. The geographical distribution of freshwater decapods and its bearing upon ancient geography. *Proceedings American Philosophical Society* 41:267-400.

O'Sullivan, D. O. S. 1991. Status of Australian aquaculture in 1989/90. *Austasia Aquaculture–June 1991/Trade Directory*:2-13.

Patak, A. 1982. A structural and immunochemical comparison of the haemocyanin of Australian Parastacidae. BSc (Honours) thesis, Department of Zoology, Monash University, Victoria, Australia.

Patak, A., and J. Baldwin. 1986. Electrophoretic and immunochemical comparisons of haemocyanins from Australian freshwater crayfish (Family Parastacidae): Phylogenetic implications. *Journal of Crustacean Biology* 4:1105-1120.

Pidgeon, R. W. 1981. Diet and growth of rainbow trout, *Salmo gairdneri* Richardson, in two streams of the New England Tableland, New South Wales. *Australian Journal Marine Freshwater Research* 32:967-874.

Reynolds, G. 1984. Predation. In Marron farming, *Proceedings of a Workshop held by the Marron Growers Association of Western Australia (Inc.)*, October 1984, Bennison, S., (Ed.), Perth, Western Australia, Australia pp. 55-57.

Reynolds, K. M. 1980. Aspects of the biology of the freshwater crayfish *Cherax destructor* in farm dams in far-eastern New South Wales. MSc thesis, School of Zoology, University of New South Wales, New South Wales, Australia.

Reynolds, L. F. 1976. Decline of the native fish species in the River Murray. *SAFIC* 8:19-24.

Riek, E. F. 1951. The freshwater crayfish (family Parastacidae) of Queensland. With an appendix describing other Australian species. *Records Australian Museum* 22:368-388.

Riek, E. F. 1956. Additions to the Australian freshwater crayfish. *Records Australian Museum* 24:1-6.

Riek, E. F. 1969. The Australian freshwater crayfish (Crustacea: Decapoda: Parastacidae), with descriptions of new species. *Australian Journal of Zoology* 17:855-918.

Riek, E. F. 1971. The freshwater crayfishes of South America. *Proceedings Biological Society Washington* 84:129-136.

Riek, E. F. 1972. The phylogeny of the Parastacidae (Crustacea: Astacoidea), with descriptions of a new genus of Australian freshwater crayfishes. *Australian Journal of Zoology* 20:369-389.

Ritchie, M. E. 1978. Circadian rhythms, and their measurement by microcomputer, in *Geocharax gracilis* (Clark 1936). BSc (Honours) thesis, Department of Zoology, University of Tasmania, Tasmania, Australia.

Roughley, T. C. 1966. *Fish and fisheries of Australia*. 2nd ed., Angus and Robertson, Sydney, Australia.

Roux, J. 1914. Ueber das Vorkommen der Gattung *Cheraps* auf der Insel Misol. (Aus den Zoolog. Ergebnissen der II. Freiburger Molukken-Expedition 1910-1912). *Zoologischer Anzeiger* 44:97-99.

Roux, J. 1933. Note sur quelques Crustaces decapodes d'eau douce provenant de l'Australie septentrionale. *Revue Suisse Zoologie* 40:343-348.

Sammy, N. 1988. Breeding biology of *Cherax quadricarinatus* in the Northern Territory. In *Proceedings of the 1st Australian Shellfish Aquaculture Conference*, 1988, Curtin University of Technology, Perth, Western Australia pp. 79-88.

Shearer, K. 1981. Carp, crayfish, and water rat. In *Waterplants of New South Wales*, Sainty, G. R. and Jacobs, S. W. L. (Eds.), Water Resources Commission, Sydney, Australia pp. 461-470.

Shipway, B. 1951. The natural history of the marron and other freshwater crayfishes of south-western Australia. Part 1. *Western Australian Naturalist* 3:7-12.

Skidmore, J. F., and I. C. Firth. 1983. Acute sensitivity of selected Australian freshwater animals to copper and zinc. *Australian Water Resources Council Technical Paper* No. 81.

Smallridge, M., R. Musgrove, and S. Allenson. 1989. Growth of juvenile yabbies fed on live food. *SAFISH* 14(2):6-8.

Smith, G. W. 1912. The freshwater crayfishes of Australia. *Proceedings Zoological Society London* 1912:144-170.

Sokol, A. 1988. Morphological variation in relation to the taxonomy of the destructor group of the genus *Cherax*. *Invertebrate Taxonomy* 2:55-79.

Spencer, B., and T. S. Hall. 1896. Crustacea. In *Report on the work of the Horn Scientific Expedition to central Australia*, Part II. Zoology, Spencer, B. Ed., Dulau and Co., London pp. 227-248.

Standiford, A. J., and J. Kuznecovs. 1988. Aquaculture of the yabbie, *Cherax destructor* Clark (Decapoda: Parastacidae): an economic evaluation. *Aquaculture and Fisheries Management* 19:325-340.

Standiford, A. J., J. Kuznecovs, and B. J. Mills. 1987. Economics of commercial aquaculture of the yabbie (*Cherax destructor*). Report Department of Fisheries, Adelaide, South Australia.

Treadwell, R., L. McKelvie, and G. B. Maguire. 1991. Freshwater crayfish. In *Profitability of selected aquacultural species*. Australian Bureau of Agricultural and Resource Economics Discussion Paper 91.11, Australian Government Publishing Service, Canberra, pp. 55-62.

Troughton, E. 1941. Australian water-rats: their origin and habits. *Australian Museum Magazine* 1941:377-381.

Unestam, T. 1975. Defense reactions and susceptibility of Australian and New Guinea freshwater crayfish to European-Crayfish-plague. *Australian Journal Experimental Biology and Medical Science* 37:237-242.

Walker, K. F. 1982. The plight of the Murray crayfish in South Australia. *Red Gum* 6:2-6.

Walker, K. F. 1983. The Murray is a floodplain. *South Australian Naturalist* 58:29-33.

Woodland, D. J. 1967. Population study of a freshwater crayfish, *Cherax albidus* Clark. PhD thesis, University of New England, New South Wales, Australia.

Woollard, P., V. J. M. Vestjens, and L. MacLean. 1978. The ecology of the eastern water rat *Hydromys chrysogaster* at Griffith, NSW: Food and feeding habits. *Australian Wildlife Research* 5:59-73.

Zeidler, W. 1982. South Australian freshwater crayfish. *South Australian Naturalist* 56:36-43.

Index

Bosmids, frozen, as juvenile food, 178
Bouchardina, 5
Branchiobdellid worms, 135,204
  removal during purging, 108
Brevoortia, 51
Broad-clawed crayfish. *See Astacus astacus*
Brood pheromone, 11
Broodstock
  feeding of, 243,269
  loss of, 121
  preexisting, 42
  sources of, 41
Buffalofish
  as crayfish bait, 50
  polyculture with crayfish, 64
Bullfrog, as crayfish predator, 62
Burn spot disease, 122-123,202-203
Burrow ditches, 23
Burrowing activity, of crayfish
  as agricultural problem, 189
  of Australian crayfish, 218,219, 220,223,225,228-229,250, 267
  control of, 189
  of North American crayfish, 14-17
Burrowing crayfish, classification of, 15
Burrows, 9
  "dry", 16
  in levees, 30,42

Cage, for purging, 109
Calcium
  storage during molting, 17-18
  water content
    alkalinity and, 25
    crayfish tolerance levels, 251-252
Callinectes sapidus, 27
  off-season processing, 91
Cambarellus, 5,7
Cambarellus shufeldtii, 13-14

Cambaridae, 5-6,189. *See also* North American crayfish
  life cycle, 9-14
  maturation, 13-14
  reproduction, 12-13
  reproductive anatomy, 8-9
Cambarus, 5,7
Cambarus affinis. See Orconectes limosus
Canada, *Orconectes* culture in, 84
Canal, recirculation, 24
Cannibalism
  by *Astacus astacus*, 174
  dietary factors in, 196
  evolutionary basis, 196
  among juveniles, 244,268
  by *Pacifastacus leniusculus*, 183
Carapace, removal from soft-shell crayfish, 113
Carassius auratus, 64,234
Carbon, nutritional, 33
Carbon:nitrogen ratio, of forage vegetation, 33,34
Carotenoids, 199
Carp
  as crayfish bait, 51
  as crayfish competitor and predator, 234
  polyculture with crayfish, 64,188
Catfish
  as crayfish predator, 234
  polyculture with crayfish, 64
Catostomus, 51
Cement glands. *See* Glair glands
Cephalothorax, size, 96
Cestodes, as crayfish parasites, 235
Chelae
  of *Astacus astacus*, 174
  of *Cherax bicarinates*, 266-267
  of male crayfish, 12
  of *Procambarus clarkii*, 20
  of *Procambarus zonangulus*, 20
Chelating agents, for frozen meat discoloration prevention, 107-108

302   *FRESHWATER CRAYFISH AQUACULTURE*

Mitochondrial DNA analysis, 83
Molting
 by cambarid crayfish, 17-19
 into Form II condition, 13
 handling implications of, 95
 induction of, 18
  with hormones, 18
  with water temperature,
   182,191-192,195
  initial, 11,174
 by soft-shell crayfish, 76
 stages, 17
 system turnover time of, 73-74
Molting crayfish, separation from
  nonmolting crayfish, 75
Molting trays, for soft-shell crayfish
  production, 78
Molt-inhibiting hormone, 18
Molt-stimulating hormone, 18
Mosquitofish
 as crayfish predator, 234
 for water boatmen control, 61
Moving tray, for egg incubation, 176
Muck, anoxic, 32
Murray cod, as crayfish predator,
  233-234
Muscles, atrophy of, 135
Muskrat, as crayfish predator, 62
Mustella vison, 62
Mutations, 82-83
Mycobacterium, 117
Myocaster coypu, 51-52,62

Narrow-clawed crayfish. *See Astacus
  leptodactylus*
Natural populations. *See* Wild
  populations
Nematode infestations, 135,235,254
Netherlands
 Orconectes virilis in, 165
 Procambarus clarkii in, 167-168
Nets
 baited, 263
 drop, 25-26
Netting, for ponds, 263,264,265,270

Noble crayfish. *See Astacus astacus*
North America
 Australian crayfish culture in, 83-89
 legal regulation of crayfish
  possession in, 87
North American crayfish, 5-89
 burrows of, 14-17
 commercially-important species, 6-7
 culture of, 19-89
  economics, 69-70,71-72
  forages and feeds, 32-40
  genetical considerations, 82-83
  harvesting, 48-60
  market development, 67-68
  polyculture, 63-65
  pond site selection
   and construction, 21-25
  pond types, 19-21
  population management, 40-48
  predators, 60-63
  production in tonnage, 8
  semi-intensive/intensive, 70,72-82
  transportation, 65-67
  water management, 25-32
 growth, 7
  history, 7
  life cycle, 9-14
  molting, 17-19
  size, 7
  taxonomy, 8-9
Northern crayfish. *See Orconectes
  virilis*
Norway
 annual crayfish production in, 168
 Astacus astacus culture in,
  160-162,181
 Pascifastacus leniusculus natural
  fisheries in, 168-169
Nosema, 126
Notemigonius crysoleucas, 64
Nursery ponds, 178,179,244
Nursery tanks, 257,259,269-270
Nutria
 as crayfish bait, 51-52
 as crayfish predator, 62

Procambarus, 5
culture of, in higher latitudes,
    85-86
maturation, 30
size, 7
Procambarus acutus, nutrition,
    197-198
Procambarous acutus acutus
heterozygosity, 83
production, by tonnage, 8
reclassification, 6n.
Procambarus alleni, 81-82
Procambarus blandingi,
    reclassification, 6n.
Procambarus clarkii
burrowing activity, 15-16
as commercially-important
    species, 607
consumer acceptance of, 40,204
culture of
    colonial cultivation system, 81,
        82
    egg incubation, 10-11
    in Europe, 189-190
    fish hatchery culture, 85
    formulated feeds, 39
    hatchery system, 77-78,81
    meat yield, 104
    in northern latitudes, 48
    oxygen requirements, 25
    population dynamics, 44-45
    production, by tonnage, 8
    stocking rates, 41,42
    summer production, 46-47
    temperature requirements, 25
    temporary ditch habitat, 23
distinguished from *Procambarus
    zonangulus*, 92-93
European consumers' acceptance
    of, 204
as European introduced species,
    167-168
form I condition, 12,20
holdover population, 43

interaction with *Procambarus
    zonangulus*, 45
as introduced species, 6,167-168
life cycle, 13-14
molting, 74
as parasitic worm host, 132
as plague fungus vector, 122
predators, 60-61,62
reproductive potential, 43
size, at maturity, 14
in Spain, 169-170,189-190
spawning time, 42
trapping of, 189
unintentional polyculture, 63
uropod swelling, 135
Procambarus zonangulus
comparison with *Procambarus
    clarkii*, 45-46
distinguished from *Procambarus
    clarkii*, 92-93
eradication, 46
form I condition, 20
hatchery systems, 81
interaction with *Procambarus
    clarkii*, 45
meat yield, 104
as newly identified species, 6
oxygen level requirements, 25
parasite/disease-resistance, 45-46
as percentage of total crayfish
    harvest, 40
predators, 61
production, by tonnage, 8
spawning time, 42
temperature requirements, 25
temporary ditch habit, 23
Processing, of crayfish, 91-115
blanching time, 100
of cooked crayfish, 100,108
of frozen crayfish, 100,108
grading, 95-99
harvesting for, 93
heat processing, 100-102
major locations for, 91

Rice
  as forage, 34,35,36,38
    in polyculture ponds, 64
    seasonal biomass changes, 331
  as pond thermal insulation, 46
Ricefield pond culture, 19,32,39
  biological oxygen demand in, 28
  drainage in, 32
  in Spain, 189,190
Rice paddies, crayfish as pests in,
    86,189
Rock lobster, legal size of, 255
Rodents, aquatic, as crayfish
    predators, 62,235,254,270
Roman Empire, crayfish culture in,
    157
Rotenone, for predaceous fish
    control, 62
Rotifers, 135,235
Russia
  crayfish culture in
    of *Astacus leptodactylus*, 187
    historical background, 158
    of *Pacifastacus leniusculus*, 167
  crayfish plague fungus in, 200
Rusty crayfish. *See Orconectes
    rusticus*

Sacks, for crayfish transport,
    65,66,67
Sacramento-San Joaquin Delta,
    Procambarus clarkii
    reproducing populations in,
    47
Sagittaria graminea platyphylla, 39
Salinity, of pond water, 25
  crayfish tolerance of,
    231,251-252,267
Salmo trutta, 234
Salt bath, for purging, 108
Samastacus, 220
Sarotherodon, 64
Scandinavia. *See also* Finland;
    Norway; Sweden

Astacus astacus annual catches in,
    160,162
Astacus astacus distribution in,
    160
crayfish culture in, history of,
    157-158
crayfish trapping in, 190,192,
    193,194
Scapulicambarus, 6
Seasonal factors
  in meat yield, 104
  in population density, 230-231
  in production, 91
Seasoning solution, 102
Seining, as harvesting method, 55-
    56,58,244
Selenium, crustaceans' requirements,
    199
Self-perpetuating nature, of crayfish
    culture, 14-15
Septicemia, 119-121
Sex ratio, of Australian crayfish,
    233,256,269
Sexual dimorphism, 174
Shelf-life, of crayfish meat, 105
Shell disease, 118,119
Shrimp, fungal diseases, 122
Signal crayfish. *See Pacifastacus
    leniusculus*
Silt, removal from crayfish, 110
Size
  large, 44
  at maturity, 11,13,14
  population density and, 44
  stocking rates and, 41
Skipjack herring, as crayfish bait, 50
Smartweed, 37,39
Snares, 255
Sodium, crayfish meat content, 102
Soft-shell crayfish
  culture of, 70-81
    annual operating costs, 80
    Bodker System, 74-75
    Culley System, 74
    depreciation charges, 79

Worms, parasitic, 132-135
  as Australian crayfish parasites,
    235,236-237
  branchiobdellid, 108,135,204
  digenetic trematode, 132-133
  spiny-headed, 133-135
Wounds
  melanization of, 118
  parasites of, 122

X-organ, 18

Yabbie. *See Cherax destructor*
Y-organs, 18

Zooplankton
  fertilization of, 86
  frozen, as feed, 244,270
Zoospore, of crayfish plague fungus,
    201-202,203-204
Zoothamnium, 127